THE ALBRECHT PAPERS
Enter Without Knocking

ABOUT THE AUTHOR

Since this fourth volume is really a companion to three previous books, these few notes are only a codicil to biographical notes set down before. Dr. William A. Albrecht died May 21, 1974. There was no reason why this gentle old man could not have lived many more years. His health was fair to good, and he knew how to take care of himself. "I know how they'll get me," he confided to the *Acres U.S.A.* publisher. "One day they'll get me out to one of those restaurants with all that synthetic food and preservatives, and I'll suffer an attack. I won't be able to get to my bathroom. They'll take me to a hospital, and they won't know what to do. I'll be a goner."

It happened that way. A group of colleagues ordered Bill Albrecht to a restaurant to honor him. In the convivial atmosphere of the hour, he made an error. He ate chemical-laced fare that triggered a metabolic problem he had been living with for years. They took him to a hospital. In a few days he belonged to the ages.

But before that tragic day, he had released to *Acres U.S.A.* his lifetime production of papers for recycled presentation to a new generation of readers. It has been and continues to be a privilege to continue the work of William A. Albrecht.

THE ALBRECHT PAPERS

Enter Without Knocking

**This is the fourth volume on the life
and work of William A. Albrecht**

by William A. Albrecht, Ph.D.

Edited by Charles Walters Jr.

THE ALBRECHT PAPERS
Enter Without Knocking

Published by Acres U.S.A.
P.O. Box 9547, Kansas City, Missouri 64133
Copyright © 1992 by Acres U.S.A.
Printed and bound in the United States of America.

ISBN: 0-911311-23-8
Library of Congress Catalog card number: 92-071281

Dedicated to the memory of
E. R. Kuck
founder of Brookside Dairy Farms
and Laboratories

CONTENTS

FOREWORD

Enter Without Knocking has meaning to those who knew Dr. William A. Albrecht best. It was a greeting—actually, "Enter without knocking, and leave the same way"—he would issue at frequent intervals until May 21, 1974, the day he passed from the scene. During the last several years in the life of this great scientist, I taped over a hundred hours of conversations that covered his life and career. During one of our last meetings he provided *Acres U.S.A.* with almost eight hundred published papers. Perhaps a hundred have been recycled in *Acres U.S.A.* Twice that many have been preserved in three earlier volumes of *The Albrecht Papers*. This is the fourth volume, and covers superior sections styled here as chapters. Each paper carries its original title. Notes in the back of the book provide sources.

William A. Albrecht had added "Emeritus" to his title when I first met him. The word was out that he had been retired because he wouldn't "go along," and there was some truth in the charge. He had been a scientific gadfly since 1916, when he came to the University of Missouri as a microbiologist on the staff of the Soils Department. He was the seventh of eight children reared on a farm in Livingston County, Illinois.

Even in retirement, Albrecht worked overtime at schooling people. He corresponded tirelessly. People with minimum training could count on this quiet gentleman to explain complicated chemistry in terms that could be understood. C. Haynes Thompson, the proprietor of a phosphate firm, couldn't understand the *Insoluble Yet Available* paper. "Calcium and magnesium are divalent elements, hence offset two hydrogens or two equivalent weights of it. Potassium is monovalent, hence offsets only one equivalent weight of hydrogen per atomic weight (not per atomic number)."

Albrecht's visitors were farmers, consultants, and students in the main, but well known people of vision also came to pay homage. One day, quite unannounced, I drove down to Columbia with Eddie Albert, the movie and TV personality. Albrecht gently excused himself from a recording session with the university archivist and gave Eddie Albert both barrels: an hour-long lecture on the Pottenger cat studies, and another on the miracle of the Ozarks. I can remember Albrecht's lesson as if I heard it last night, and I spent the trip back to

ix

Kansas City visiting with Eddie Albert about what Albrecht had said.

Lady Eve Balfour of England found a lot of news when she heard Albrecht intone, "Enter without knocking, and leave the same way." She was making a tour of the United States to learn about the state of agriculture in what the British once called "the colonies," and what she learned both alarmed and delighted her. The highlight of her trip was a study session with Bill Albrecht. You didn't just visit the professor, you studied. For Lady Balfour, Albrecht was in the Weston Price-Sir Robert McCarrison class. The literature, of course, was available to Lady Eve Balfour, and in her report she assembled some memorable statements. But it was Albrecht's critique of the ash mentality that captured her imagination, and she returned to England the best of the best from the U.S.A.

"When the effects from fertilizers on soils are measured only by yield variations in vegetative bulk, recorded in tons and bushels, there is little chance that we shall recognize crop differences demonstrating the varying effects between the use of inorganic and organic fertilizers. Our animals, however, tell us that the crop's nutritional quality reflects the different organic and inorganic compounds feeding the plants. When we learn to measure the crop's responses to soil fertility by more than bulk values and ash differences, then the contributions of the soil, both organic and inorganic, to plant nutrition will be more correctly realized," Albrecht said.

One day when I came to Columbia to see Albrecht, he was moving out of his office in Mumford Hall. His library was for sale. I filled my trunk with books and enjoyed a fine lunch with the professor and his wife, Gertrude. Bill Albrecht could no longer drive, but his mind was still as sharp as a glass bell. He apologized for warming slowly, and he admitted he just couldn't trust economists, not even Thorstein Veblen, who used to teach at the University of Missouri, Columbia. In fact, he had always voted to keep that discipline out of the professional societies to which he belonged.

It was the last time I would hear him intone, "Enter without knocking, and leave the same way."

This volume continues the news reported in the first three volumes.

—*Charles Walters Jr.*
Publisher/Editor, Acres U.S.A.

1

THE NITROGEN FACTOR

IS NITROGEN GOING WEST?

The French farmer collects all vegetable wastes about the farm and brings in forest leaves, bracken fern and any wayside or swamp growths that represent possibilities for animal bedding and more manure. Barn manures, supplemented by any farm refuses, are composted into the carefully built, flat-topped, and straight-sided pile for the heating or fermentation process by which the European farmer truly "makes manure." This finished product is then put on the field in regular, seemingly measured piles from the two-wheeled ox-cart or single-horse wagon. Later it is spread very carefully at a uniform but relatively heavy rate. When one observes the laboriousness of this procedure and the detailed attention to farm manure in contrast to the way we, in the United States, hastily get rid of barn wastes with the minimum of thought except for their speediest disposition, it provokes thought about nitrogen, and other fertility elements, so parsimoniously treasured by the farmer in Europe. It reminds one that the organic matter, and in it the nitrogen, of our own soils in the United States has seemingly taken to following the advice of Horace Greeley, and it too, like many other things, is "going west."

Viewed in relation to the map of climate of the United States, and consequently to the forces developing our soils, the eastern states under higher rainfall have a low supply of soil nitrogen in relation to the carbon, or they have an organic matter of a wide carbon-nitrogen ratio. The western states have the reverse, namely, a high supply of nitrogen in the microbial diet of decay as related to the carbon, or a narrow ratio. Then, too, in the former area, which was originally one of forests, the carbon was added on the top of the surface soil by the annual drop of the highly carbonaceous, tannin-laden leaves. In the latter area, the addition was one well distributed through the surface soil in the form of the annual contribution of deeply penetrating, more proteinaceous roots of native grasses and legumes. These features, carefully considered, point to an interesting pattern of varying carbon-nitrogen ratios along any line of constant temperatures but of decreasing rainfall as one goes west.

Beginning in the eastern United States, one's going west means a shift from conifers to deciduous forests; a shift from soft woods to hard woods. In going still farther west there is a shift from hard wood forests to starchy food grains; and then, even within the wheat grain area, a traverse westward as limited as one across Missouri and Kansas means a shift from soft wheat to hard wheat. From "soft" to "hard" in the case of wood means more ash as to both amounts and kinds of chemical elements included. From "soft" to "hard" in the case of wheat means also more ash and chemical elements in the products of plant growth. In both the wood and the wheat, there has always been an increasing nitrogen content connected with the increasing concentrations of the chemical ash elements or the fertility in the plants as taken from the soil. In the wood, however, we have emphasized the ash and disregarded the nitrogen. In the wheat our emphasis has been reversed. We have pointed to the nitrogen as protein, making the so-called "hard" wheat. But we have disregarded the high ash or fertility taken from the soil that plays an important part among the soil conditions making the wheat hard in terms of its synthesized protein at the same time that they make it high in fertility ash.

Harder wheat represents a greater microbial "turnover" annually of the nitrogen in the soil organic matter which gives more of the soluble nitrogen. It means, too, that there is no deficiency in the microbial diet of the mineral nutrients in relation to its supply of energy. Microbial "burning" of the carbon within the soil puts the carbon out of the soil as the gas, carbon dioxide. It gives a narrow carbon-nitrogen ratio in the humus of the soil. It brings the nitrogen out of its less soluble, organic combinations and puts it into the more soluble, mineral, nitrate forms. This plant nutrient is thereby no longer in the hoard of the microbes which are competing with the crop for it. It is released and can move downward in the soil to be temporarily recombined, or to remain soluble in the deeper soil horizons as one finds it more commonly in the soils of the western, hard-wheat belt where the lower annual rainfalls do not leach it down and out through the soil profile.

Under the higher annual rainfalls the soil situation represents a shortage of mobile nitrogen in terms of the crop, both because of the leaching of the minerals and nitrates, and because of the control of the nitrogen by the microbes that are keeping it combined with the excessive carbon. The same situation prevails in old agricultural soils farmed for many years. There the microbial fires had been fanned so long by tillage as to have burned the humus out and lost the nitrogen. Only a woody crop growth is now possible there as green manure for organic matter returned to the soil. There some nitrogen can be released in the form of mobile nitrates in the limited surface soil only during the spring by tillage and the rising temperatures that burn the carbon and reduce the carbon-nitrogen ratio. But that nitrate supply is soon exhausted by the multiplying roots of the wheat crop, for example, which spends this protein-producing nutrient in making its vegetative skeletal structure. Over the usually infertile subsoil, the nitrogen is then deficient in the plant's later, seed-making performances which are limited to putting out mainly starch and are prohibited from converting much of that into the liberal content of protein that makes a hard wheat. It is not, therefore, the high rainfall of the eastern states that makes soft wheat, but instead it is the limited supply of soluble nitrogen in the soil at the time when the plant is really "making the grain," as the farmer commonly describes it. More mobile nitrogen in the soil during a particular phase in the crop's physiological cycle comes about through a relatively limited carbon supply as energy for the soil microbes. It is such conditions that allow the nitrogen to occur as mineral form or nitrate. It is these conditions that we provide, for example, when we turn under nitrogenous green manures, or as we feed the microbes on a diet with a narrow carbon-nitrogen ratio so that the nitrogen can be released from the carbon and move downward in the soil. It is in such that we can see the wheat crop prompted to send its roots down for their arrival at this more abundant supply of soluble nitrogen so timed as to deliver this nutrient, not early and for excessive vegetative production, but late in the plant's growth cycle for conversion of the starch into protein and for making the hard, high-protein wheat grain.

Such hard wheat has commonly been believed to be caused by the low rainfall rather than by the complex set of soil conditions including: (1) a liberal supply of mineral fertility to encourage microbial conversion of organic nitrogen into nitrates; (2) a limited annual rainfall that prohibits excessive vegetative growth and does not seriously leach the soluble nutrients out of the soil; (3) the narrower carbon-nitrogen ratio of the organic matter grown and decayed there; and (4) the location of the nitrates in liberal quantities in the deeper soil horizons to provide nitrogen generously those stages in the plant's physiological processes that may mean less yields as bushels per acre but in place of a soft, starch, or highly carbonaceous grain they mean more in terms of a highly proteinaceous food product of the quality so much desired.

The exploitation of the soil fertility has compelled the hard wheat to go west. That condition of the soil is inducement also for other phases of agriculture to go west. Is it pure coincidence that Kansas City surpassed Chicago recently in the number of beef cattle handled annually on the market? Isn't this a case where our big protein crop, or meat, is already in the West? There is the suggestion, too, that hard wheat is going still farther west, if the complaints by the bread bakers in Kansas are an indication. They say that the Kansas farmers are abandoning the varieties of hard wheat and taking to those of soft wheat. They claim it is high time to search and research for new varieties of hard wheat. The farmer will, of course, grow whatever kind of wheat maintains his volume of output as measured in bushels per acre. For him that volume may as well be a soft wheat as a hard wheat when volume alone or bushels per acre determines his earnings. Bakers want to maintain their volume of output, too. But this is of a bread loaf that can be blown into big volume from the minimum of flour and with maximum retention of water as guarantee of weight. This is possible only by means of protein or colloidal nitrogen put into the dough-mix by way of some particular soil conditions under the growing grain. Soft wheats of low protein, or low nitrogen content can't meet the baker's specifications even though they may well satisfy those laid down by the farmers. Shall we, therefore, match the bakers against the farmers in a squabble about varieties of wheat, or shall we suggest that both groups have a common or mutual problem in the dwindling fertility of the soil?

Both the bakers and the farmers may well be reminded that hard wheat is "going west" for them as it did for the folks in the Genesee River Valley of New York State where hard wheat once made that area the "bread basket" of the United States, and Rochester, "the Flour City." When they can't move on or go west with their wheat-growing and baking businesses, shall they plant to reduce their outputs or shall they think about using some nitrogen as fertilizer on the soil? For the recent five years, 1941-1945 inclusive, Kansas has had more favorable rainfall and has produced wheat yields each year approaching the state's maximum crop of 1931. These five years represent a rate of nitrogen removal from the soil in the wheat crop almost double that of a previous five-year period, like 1932-1936 inclusive, for example. Cannot such excessive removal of nitrogen and other soil fertility elements from the soil by these large crops be considered as casually connected with the declining quality of the wheat? Is it not wiser that the bakers and farmers should recognize the dwindling soil fertility as a mutual problem rather than that they should fly at each other with accusations of the one attempting to maintain his own high volume of output go to the other? Would it not also be a wiser national policy in terms of food values in our "staff of life" to think about keeping high the protein and ash contents of the wheat by fertilizing the soil than to agree to lower the standard figure of percentage of protein for a product that all of us eat?

Yes, I believe you will agree that nitrogen is "going west," at least in the proverbial slang sense of that term, when we merely shrug our shoulders with indifference to the conservation of our soils. But if we are going to have a hard wheat always and a highly proteinaceous flour, nitrogen may be going west in a distinctly geographical sense and in terms of its application more widely as a fertilizer in our better attention to the fertility of the soil. We shall then be treasuring the chemical nitrogen of industrial output as well as the urinary nitrogen of farm manures which the so-called "French peasant" is still guarding today as he always has—ever since the revolutionary days when nitrogen from his manure pile meant gun powder in the fight for freedom from dictators.

IS COMMERCIAL UREA AN "ORGANIC" FERTILIZER?

Recently an organic gardener raised the query, "Is not the commercial nitrogen fertilizer, urea—which is chemically made from nitrogen of the air—improperly called an "organic fertilizer?" This is not a recent question. Urea had its discovery as a white crystalline compound in urine in 1773 by Rouelle. As a waste product of the body's breakdown of proteins, it originated in a living organism and was considered organic for 55 years. Then, in 1828, a German chemist, Wohler, made urea from some laboratory chemicals, potassium cyanate and ammonium sulfate. Since that date, folks could take their choice as to whether they want urea to be called an "organic" or an "inorganic" compound, so far as its origin is concerned.

In considering commercial urea as a fertilizer, (in contrast to it in manure) it is well to be reminded that the chemical urea compound has long been used, and in considerable quantities, when given off by another inorganic substance, namely, calcium cyamid containing nitrogen of atmospheric origin. When put into moist soil, this nitrogen fertilizer breaks down liberating urea slowly. The change may take a week. In some conditions, intermediate or secondary compounds resulting may be disturbing to seedling growth. But this reaction is transient and its dangers can be prevented by making the soil treatment with it ahead of the planting. Here, then, is another fertilizer combining air nitrogen by means of limestone and coke to make an "organic" fertilizer, urea, result within the soil, and one no different than the urea from animal urine. In terms of more common chemical thinking, however, urea should be considered an "organic" substance, since by its composition of carbon, hydrogen, oxygen and nitrogen it leaves no ash on burning. Also, the nitrogen has two hydrogens connected with it and is itself, in turn, attached to a carbon.

This is a characteristic chemical structural arrangement of the nitrogen in proteins, the living compounds of carbon, hydrogen, and nitrogen (sometimes also of sulfur and phosphorus) which can grow, protect themselves from other proteins, and reproduce. In the case of urea, this arrangement of nitrogen is called the "amide" of carbonic acid (organic origin) because of the nitrogen-hydrogen combination in a larger compound.

Used as a soil treatment, or as "activator" in the compost heap to balance the microbial diet, the behavior and effects of the urea are the same regardless of its origin in an animal or in the manufacturing laboratory.

It changes into ammonia, water, and carbon dioxide quickly, especially under the influence of microbes or even enzymes in some plant products. The ammonia released may be held by adsorption on the soil colloid, may be taken as such by microbes or plants for their nutrition, or may be converted into nitrites and then nitrates for more nutritional services by those forms.

Of course, urea is not the equal in effects by urine in composting straw and other fibrous wastes, when the latter carries both other inorganic and organic compounds not present in the chemically pure urea. So in terms of nutrition of the plant and of the microbes these life forms draw no distinction as to whether you call urea an inorganic or an organic fertilizer according to its origin. For them this compound is one and the same in function, regardless of our possible quibble over its particular classification into the one or the other of these categories.

AN OLD PROBLEM—LOSS OF APPLIED NITROGEN

"The second quarter of the 19th Century saw the introduction to European agriculture of guano, nitrate of soda, and sulfate of ammonia. The results produced by the first of these were in many instances little short of marvelous and, accordingly, the popularity of guano grew very rapidly in England and on the continent."—J.G. Lipman and A.W. Blair in their first paragraph of New Jersey Experiment Station Bulletin No. 288 (January 4, 1916), which discusses their "investigations relative to the use of nitrogenous plant foods," which were begun in 1898.

In their second paragraph we read, "The early use of nitrate of soda and of sulfate of ammonia was, on the other hand, accompanied by far greater uncertainty." Then quoting from a book on agricultural chemistry published in London in 1849, "Nitrate of soda, therefore, must be considered as one of those partial manures upon which no safe reliance can be placed, the application of which has caused so much perplexity, disappointment and loss to farmers."[1]

"Similarly," their report continues, "the results from the use of ammonium sulfate were at times quite unsatisfactory, and instances were noted where the application of ammonium salts proved injurious." This mid-nineteenth-century report told us of the more efficient nitrogen for plant nourishment in guano—bird droppings from the dry western coast of South America—and also of the nitrogen from the salts of guano, namely nitrate of soda and sulfate of ammonia. Farmers who tested both forms of nitrogen spoke up more vociferously for the bird manure than for the chemical salts.

Lipman and Blair report on the two camps, or factions, speaking for the two kinds of nitrogen. "Liebig and his followers stoutly maintained that the farmers could produce abundant crops without the purchase of ammonia (nitrogen). "You must learn," he told the farmers, "to draw as much nitrogen from natural sources as is necessary for your purpose. Thousands of facts teach us that this is possible."[2] On the other side, those whom Liebig called "The Nitrogen School" had Lawes of England asserting that "the increased production of grain depends on a liberal supply of ammonium salts," and that "the yield of crops bore a direct relation to the amount applied." Today we have a similar situation. Now that nitrogen in its many forms (gas, solutions and salts) has been used as fertilizer more extensively than ever, the efficiency with which those forms are utilized by the various crops from the larger amounts applied to the soil is a question again under research at many experiment stations.

The first of these conflicts, occurring in the middle of the 19th Century, stimulated Lipman and Blair to make tests and report on "The Use of Nitrogenous Plant Foods." The recent conflict has led to duplication of the tests which these men made at the New Jersey station, rather than to a study of the unopened publications of New Jersey for knowledge of the efficiency of nitrogenous fertilizers in customary agriculture. In these early experiments, steel cylinders, open at top and bottom, were buried. Each was filled with 175 pounds of a uniform lot of virtually barren soil containing almost no nitrogen. The treatments stepped up from the "check" to include quite regularly the "minerals" to supply phosphorus and potassium and some lime; then those elements plus manure (first solid only, then solid and liquid, fresh and leached), nitrate of soda, sulfate of ammonia, dried blood and manure plus nitrate. Crop rotations were used in each cylinder. The study lasted 15 years and involved 21 crops. These experiments showed that "the various nitrogenous materials which were studied influenced, each in its own way, not only the chemical but also the physical and biological properties of the different soils, and that the changes so produced have reacted markedly on the crop yields. It has been found that the system of cropping followed involved a steady loss of nitrogen, even where applications of cow manure at the rate of 16 tons per acre have been made annually. It has been shown how the composition of animal manure and its previous treatment may affect the yield and composition of the crop grown." Of the series of 20 treatments, four of the five yielding the maximum of nitrogen (crude protein) during the 15 years included the use of manure plus nitrate nitrogen, and the fifth used dried blood instead of the nitrate salt. The five treatments yielding the maximum of nitrogen (crude protein) during the 15 years included the use of manure plus nitrate nitrogen and the fifth used dried blood instead of the nitrate salt. The five treatments yielding the minimum of nitrogen were (a) minerals, manure and ammonium sulfate, (b) minerals and dried blood, (c) minerals and nitrate of soda, (d) minerals only and (e) check.

As the available nitrogen only was increased in the soil, there was a higher proportion of it in the crops. There was also "a slightly higher percentage of nitrogen in the dry matter." The highest mean of the crops during 20 treatments was 1.1444% of nitrogen in the dry matter. The lowest mean was 0.909%. Also the average percentage in the check where no nitrogen was applied was higher than in some of the series where manure and nitrate were used." As for utilization of the nitrogen applied to the crop, "The average percent of nitrogen recovered . . . ranged from 64.15 to 23.30. We must realize something of the enormous loss that results from the present methods of using (nitrogenous) fertilizers." Even though we are now using gaseous nitrogen and the urea form of it, we scarcely need to be remeasuring the efficiency of applied nitrogen at the many experiment stations.

It is significant to note that with the nitrogen applied to the soil "the average loss for the last five years was less than half the amount for the first 10 years." Our soils now do not have so much nitrogen in them and consequently there is not so much to lose. Nearly 50 years ago Lipman and Blair told us through their experiments, wherein the crops were regularly removed, in their geo-climatic setting, "It is not possible to maintain the nitrogen supply of the cultivated soil through the use of commercial nitrogenous materials alone when applied in ordinary amounts. Even with liberal applications of manure and nitrogenous fertilizers over a period of 15 years, the nitrogen content under constant cultivation was not maintained." Since protein is a nitrogenous compound, the experiments of Lipman and Blair indicate a decline in quantity (and most likely in quality) of protein when we consider nutrition from the soil as well as from added organic bulk. This fact has been established—but not widely recognized—for nearly half a century.

NITRATES . . . POSSIBLE POISON GROWN INTO FOODS

The conversion of organic nitrogen of the soil into the highly oxidized form of nitrate (saltpeter) is a natural microbial activity in any fertile soil. This soluble kind of nitrogen is taken readily from the soil for normal plant nourishment. But under disturbed conditions for the plant and for the soil microbes, like drought and its disruptions, these may cause the nitrates to accumulate within the plant rather than to function normally for their synthesis into plant proteins. Excessive nitrates in forages and foods have brought disturbing publicity in the last few years. That was a sequel to the shortages of soil moisture in the mid-continent for the highly-fertilized corn crop in particular.[1] The damaging effects of the high temperature on the plant's physiology have not been discussed. Emphasis has gone to the dangers from nitrogen applied as a fertilizer. There have been animal deaths on some soils not given fertilizers and such even before the wider use of them as soil treatments.

It is extremely difficult to prove any substance poisonous in feed or food, especially when it comes from a fertilizer—by way of the soil and the crop plants—to animals and man. For such proof in court, it seems almost necessary to exhibit the corpse, or some morbidly clinical case. A sub-clinical one is not sufficient. Animals killed by nitrate poisoning have been periodic exhibitions. They represented terrific losses to individual, but scattered, farmers over many years. They have been reported in the publications of the Experiment Stations in certain climatic sections of the United States and with careful diagnoses as far back as 1895.[2] Potassium nitrate (saltpeter) was found crystallized out on drought-stricken corn stalks which had poisoned cattle eating it with the fodder. The burning of the dried vegetable matter gave a sparkling effect of miniature explosions of gunpowder. That phenomenon was considered a simple test for nitrates as poisonous feed.

The pioneers, who used nitrates in the salt mixture for tanning pelts and hides, made certain they were kept away from livestock, which were inclined to lick or chew the salty skins with lethal results. In the first decade of this century studies were under way at the University of Illinois which used humans in so-called "poison squads" to test the effects of saltpeter "cure" of pork on men consuming such in contrast to those taking similar meat cured without nitrates. Nitrates in different vegetables have also been under study. They, like the forages for animals, proved that the soil, modified by the variation in the weather during the growing season or by certain soil treatments, could be a disturbing factor responsible for wide variations of the concentration of nitrate nitrogen in the vegetative mass. Naturally, the concentrations were low in the seed parts of the plants.

Professor J.K. Wilson, a distinguished (deceased) microbiologist of the Cornell Experiment Station, who was interested in the biochemical transformations of soil organic matter, reported many determinations of nitrates in plant juices, extracts and dry matter as early as 1942.[3] His observations deserve attention anew, buried as they are now in unopened volumes. Wilson's report revived interest in an earlier publication of the Kansas Station. In that,[2] N.S. Mayo reported on cattle killed by eating drought-stricken corn fodder in 1895. That grain crop was then just going west, in its move from the humid to the semi-humid area. C.S. Gilbert, et al., of Wyoming, reported nitrate poisoning for cattle by weeds under drought effects in 1946.[4] Two years later, the Easter Sunday edition of the *Kansas City Star* had the headline, Win in Dash for Life, on March 28, 1949, reporting the recovery of a baby from nitrite poisoning via shallow well water used in its formula, after a dash to a Kansas City hospital.

In the Kansas situation, a very dry autumn and early winter were followed by an accumulation of later winter snows. Those were melted by the early spring rains to move the soil nitrates into the wells of mid-Kansas, where only the third one of the reported babies so poisoned was saved from death. It was those buried ideas in the older reports that prompted Professor Wilson's extensive testing of leafy vegetables, frozen foods, baby foods, and other possible nitrate carriers among foods to learn whether they may contribute to hemoglobinemia, or other toxic, if not lethal, conditions for adults. Some of his data for vegetables are given in the accompanying tables 1, 2, and 3.

Those earlier reports tell us that, "When the nitrate is ingested, it passes into an environment where oxygen is in demand; the nitrate is reduced to nitrite; and the latter is toxic. This nitrite, once in the blood stream, combines with the blood. It produces a methemoglobin to prevent the blood from supplying oxygen to the tissues, and thus the body suffocates . . . The disease was reported similar to poisoning by hydrocyanic acid."

TABLE 1

Nitrates in green vegetables collected August-September 1942 (As KNO_3 parts per million in expressed juices)

Beets	1,333
Broccoli	785
Cabbage	2,222
Celery	1,212
Cauliflower	2,000
Carrots	442
Cucumber vine	4,170
Cucumber fruit from above	121
Lettuce (head)	1,250
Lettuce (curly)	870
Lettuce	666
Water melon vine	10,000
Water fruit from above	500
Water vine	1,000
Water fruit from above	1,000
Tomato fruit	0

TABLE 2

Nitrates in juices of green vegetables on market May 1948 (KNO₃ equivalent ppm)

Asparagus	50
Cabbage	1,200
Cabbage	186
Carrots	325
Celery	1,240
Celery (pascal)	3,232
Lettuce (curly)	1,818
Lettuce (head)	400
Potatoes (new)	63
Spinach Plant 1	2,353
Spinach Plant 2	1,600
Spinach Plant 3	2,000
Spinach (quick frozen)	3,636

TABLE 3

Nitrates in some baby foods on the market (KNO₃ equivalent in ppm)

Green beans	37
Beets	333, 750
Carrots	3, 0
Peas	0
Spinach	616, 833, 0, 0
Tomatoes	0
Vegetable soup	37

Wilson cited the report that 1.5% of potassium nitrate in the dry matter of feeds was known to be lethal to livestock, or to cause abortion. Missouri more recently cites that amount as deadly; reproduction difficulties at 1.0-1.5%; symptoms suggesting vitamin A deficiency at 0.6-1.0%; and no trouble with less nitrates if a normal ration is fed. For the concentration of nitrate nitrogen in water for babies, J.G. Heart suggested the maximum safety limit of 10 ppm.[5] We are a bit slow to learn that the soil is a biochemical matter about which we should be concerned, lest we upset nature as we try to manage her efforts more for our economic advantages in labor and monetary outlay. Hidden ideas in the past can help us see more of the natural in the garden and the field, if we will but open to some of these basic facts in the recorded experiences escaping us because the books remain unopened.

SOIL NITROGEN STILL A BIG PROBLEM

Among the many soil problems that have come on the stage to hold public attention, perhaps that of maintaining nitrogen has been the one factor most persistently before us. Though soil erosion has held a prominent place in our agricultural thought during the last decade, that does not mean that the decreasing nitrogen in the soil has not also played for attention right along with it. Our mental gaze needs to be fixed on the deficient nitrogen, while this may be prompting the erosion, since both are playing villainous roles simultaneously on the same gardening stage. We need to see not only the erosion, but also the declining soil nitrogen. It is this that has reduced the soil's capacity to grow its own cover crop quickly.

As combinations of several different amino acids, proteins suggest different possible compositions varying widely according as the plants are of different species and as they are nourished differently by the soil. For example, by using different ratios of the potassium to the calcium—both in exchangeable forms on the colloidal clay—Dr. Graham reduced the crop yield of soybean forage by 25%, but yet synthesized in it more protein as measured by the total nitrogen. And yet he increased the concentration of phosphorus by 100% and calcium by 200% through the modification of only the ratio of these two fertility elements. Can we feel that the physiology of the soybean plants was a constant behavior and that it was building the same kind of protein in the larger as it was in the smaller soybean crop? The need to differentiate between the kinds of proteins synthesized by the plants under varying ratios of the elements of fertility and under the variable supplies of different items of fertility in the soil poses one of the nitrogen problems that is high on the research agenda. This is especially true if we are to shift the protein-providing responsibility in our diet away from meats and milk and more toward the foods of distinctly vegetable origin, and thereby a step closer to the soil fertility as the foundation of the entire biotic pyramid.

Plants are not so directly synthesizers of the amino acids and their combinations but rather indirectly via their carbohydrates. The biosynthesis, or synthesis by the plants probably through the use of stored chemical energy, of the proteins is very probably a sequel only to the photosynthesis of the carbohydrates of sunshine energy. Then, too, plants are providers of calcium, phosphorus, and other nutrient elements. These elements are the soil fertility that controls not only the photosynthesis and the biosynthesis, but also, apparently, the construction of the catalyzing agents, like the vitamins, hormones, etc., that are provided by the plant for us as essentials along with other compounds classified more distinctly as foods.

Dr. S.H. Wittwer has recently shown that the soil fertility, particularly the nitrogen, is closely connected with the elaboration of some of the vitamins in spinach, for example. If these catalyzers which we call vitamins are connected with the nitrogen as soil fertility, this is another problem for soil research that will connect the soil with nutrition of the animals and humans through a very fine chemical thread.

The efficient utilization and elaboration of nitrogen by plants, both legumes and non-legumes, calls for a liberal supply of calcium, phosphorus, and other mineral fertility coming from the soil, as well as for the necessary supply of nitrogen. The physiological demand for calcium by the plants has been almost blanketed out and put into lesser significance in our thinking by believing that lime was beneficial for crops only through its reduction of the degrees of soil acidity. Proteins, especially such as meats, eggs and milk, have long been considered a major problem in the struggle for food.

THE ROLE OF NITROGEN

The widespread use of legume crops is premised on confidence that their growth is a help in building up the soil. That this improvement occurs merely because of the growth of the crop is the belief of many. There is a rather common inference that since legume crops can take nitrogen from the air, they need none from the soil. There is, also the erroneous inference that it will seriously disturb them if any fertilizer nitrogen is used on the soil. Legume plants are not 100% efficient in using nitrogen from the air. They do not "fix" all of the nitrogen contained in the final crop. Fixation is an accomplishment that does not necessarily start with the germination of the seed. The unique process demands the plant root's association with nodule-producing bacteria, and the nodules in larger numbers cannot be present until the root system has become extensive. Like all other plants, either non-legumes or legumes, the young legume plant is dependent on the nutrient elements stored in the seed and those which it can take from the soil. Nitrogen cannot be omitted from this list.

Nitrogen is taken from the soil supply by young legume plants in the same manner as it is used by non-legume species. Legume plants get off to a better start and subsequently fix more nitrogen when they get this initial fertility help. On a nitrogen-free soil the early life of the legume is a nitrogen-starvation period. This delays the advent of, and likewise shortens, the season of nitrogen-fixing activities. The stunted crop can deliver neither bulk nor service in fixing atmospheric nitrogen comparable to the good start of growth it can make through the help of some available nitrogen in the soil. This early use of nitrogen by the root of the legumes may possibly be a factor in the greater effectiveness with which it mobilized other soil-borne nutrients through its semi-permeable membranous cover for plant feeding.

It was recently discovered that the roots of inoculated soybeans, that were richer in protein or nitrogen as a consequence of nodulation, were able to take nutrients off the colloidal clay of the soil to a higher degree than were the roots of the uninoculated plants. As a substitute for the effect of the nodules in the plant's later life, it seems possible, therefore, that a little help by nitrogen in the soil early in the plant's life might facilitate its better use of those mineral nutrients in which it is so much more concentrated and upon which it is so dependent for efficient growth. It was Giobel of New Jersey who pointed to the value of some "starter" nitrogen for legumes, but his suggestions have not yet become farmer experience. There may be something to be gained by employing fertilizer nitrogen to help the plants get off to a good start through more efficient utilization of the mineral nutrients of the soil.

Now that erosion has given us many acres of exposed subsoil and abandoned lands for reclamation, the question may well be raised whether only lime, phosphate, legumes and livestock constitute the entire restorative procedure. Demonstrations are suggesting that legumes can survive even at such low fertility levels. But whether they can build up the soil where it will make a grass sod with its protective benefits at appreciable rates seems doubtful. Naturally, then, there is the suggestion that on such lands that economically do not permit regular and extensive use of nitrogen as a direct fertilizer for grass, we may well give some thought to nitrogen as a starter for the legumes and, indirectly through them, for the grass. The striking benefit to legumes from as little nitrogen as is contained in animal droppings prompts such a suggestion. Legumes can be soil builders, but for that service they require some foundation on which to build and some help in the building process. One cannot lift himself simply by pulling on the boot straps. Neither can legumes build much nitrogen at the outset.

As our soils are less fertile in this element, as well as in calcium, phosphorus, potassium, and other nutrients, can we expect legumes to restore them when, as a general farm experience, we know that this plant family gives us best results on the soils more fertile in all respects? There are still some questions to be answered before we can believe that legumes can build nitrogen into our less fertile soils, and even those more fertile in respect to minerals, without some help from nitrogen from other sources than the abundant atmospheric supply. There may be wisdom in the idea of helping nitrogen-fixing crops by undergirding them with available nitrogen in those soils on which we hope for them to exercise their uplift.

SOILS NEED 'LIVING' FERTILITY!

Since proteins are the only organic compounds that carry life, all life can live only because it either appropriates or synthesizes proteins required as a particular set of amino acids. Each of those acids, in turn, is synthesized from elements—carbon, hydrogen, oxygen, nitrogen, phosphorus and sulfur by only two life forms—plants and microbes.

Microbes and plants are protein-creators because of their closer connection with soil. They are protein synthesizers by means of creative power within the handful of dust. That power is fertility of soil.

Science still leaves us far short of complete comprehension of all biological and biochemical facts of nature's growing of crops and livestock.

Our sciences of agriculture depend much on only inorganic chemistry. We study ash of plants and inorganic elements of soil. By matching those two quantitatively against each other we learn but little, and that only as one or more of the required nutrient elements are absent from soil, and thereby from plants. Even when we use modern concepts of prevailing chemical tests of soil, we do not gain much assurance about soil needs—its treatments required to grow proteins, and those nutrients essential in animal nutrition. Through chemical test, we have only broadened our concepts somewhat by our improved understanding of how fertility of soil must be insoluble but yet available to the microbes and plants during the growing season. About soil tests for organic matter (nitrogen, carbon, phosphorus, and sulfur), we are still much in the dark. Sulfur, carbon and nitrogen are lost to air on converting plant to ash. Consequently, we fail to believe that organic compounds of soil are taken by plants as nourishment. We have left out of consideration, as an unknown, the soil's organic part or what seems to be half or more of the creative power of soil.

You might not consider phosphorus an organic part of plant nutrition, yet it is from the rancher's belief that phosphorus should be mixed and applied with barnyard manure, that research attention is now testing and establishing biochemical wisdom in that theory. Studies in the greenhouse and soils laboratory, by means of radioactive phosphorus grown into barley and that used as dried green manure for a succeeding crop of soybeans, have shown phosphorus in that organic form 100 times more efficient in feeding itself to the soybean crop than was true for other forms of phosphorus in a soil which was high in that nutrient according to soil tests. Sulfur, a key element in normal processes of cell division, is also related to organic matter of soil. Up to this date, we do not list sulfur as an inorganic element under fertilizer inspection. That service, emphasizing solubilities in water as its criteria, lists only nitrogen, phosphorus and potassium for certification of fertilizer's contents.

Nitrogen, measured by ignition or ashing in sulfuric acid and taken as the index of protein in crops, represents gross distortion of natural facts, when the claim is made that by fertilizing the soil with nitrogen to feed the soil microbes and crops, one gets higher concentrations of proteins grown into plants on that soil.

It has required some devastating drouths with deaths of animals by high nitrate content in the crop, like corn, to convince many that by measuring total nitrogen in corn plants and multiplying that by 6.25 and calling the result protein, we are not measuring living proteins.

Nitrogen as fertilizers for non-legume crops, like corn, may be a dangerous disturbance to nutrition of animals grazing them, especially, after dry spells. Also for legume crops, nitrogen fertilizers have demonstrated refusal of hay in trials by rabbits offered in different lots to permit their choice.

Test animals ate less hay as there was more nitrogen applied and as crop yield was larger. Such tests confirms the cow's refusal to eat tall green grass of spots in the pasture occasioned by her urinary or droppings. The animal does not measure nitrogen and multiply that result by an arithmetic factor and call it protein. The dumb brute applies the simple test of feeding value by eating it or refusing to eat it.

In interpretation of laboratory tests we have assumed too much when we believe that all organic nitrogen is protein, or inorganic elements in the ash are the only form of nutrients taken by plants. But slowly we are suspecting that organic matter is a neglected half of soil as plant nutrition and animal nourishment. Our laboratory concepts may be much farther from the facts of nature than we realize. Any crop at its best in nature, or what we call a natural climax crop, occurs only after it has grown in the same place continuously and all the successive crops have died there to build up the soil organic matter. Our removal of the annual crop of organic matter in our ranching practices violates the principle of ever arriving at a climax crop in terms of a pure stand, an excellent growth and yield, a self-protection against pests and diseases, and a fecund reproduction in terms of generous seed yields.

Hence, our neglect of maintenance of soil organic matter may be the major, but unrecognized, factor in the poor health and poor nutritional value of the crops as feed and food for animals. We may look to nature and note limitations of various forms of life to soil areas according to fertility.

Within plants, as in animals and humans, proteins must be in balance with carbohydrates which give only calories in life processes. Carbohydrates within plants serve as "starter" compounds for synthesis of proteins from them provided the balanced fertility of soil serves accordingly. Plants can convert carbohydrates into proteins, but animals cannot. Strangely enough, animals can reverse that process. Synthesis of proteins by plants seems to be a case in which some of the carbohydrates serve as raw materials out of which proteins are made. This is brought about by combining with those carbohydrates some nitrogen, some phosphorus, and some sulfur, all coming from soil. Some calcium, some magnesium, and several other soil-borne nutrient elements are required in the enzymatic agencies or other helps, while more of the carbohydrates are consumed as energy materials for bringing about this conversion process. A natural law concerning relation of fertility tells us that on soils being developed under the limited climatic forces of moderate rainfalls and temperatures—soils with a wide calcium-potassium ratio—mineral-rich, proteinaceous crops or foods as well as the carbonaceous ones can be grown.

But on soils under destruction by excessive climatic forces—those with a narrow calcium-potassium ratio—protein production is not so common while the production mainly of carbohydrates is almost universal. In the climatic setting of the former soils, nature grows healthy animals; in that of the latter soils, she fattens the less healthy ones.

Now that soils have been under cultivation for longer times, many kinds of crops have been grown on them with varied success, and animals fed the crop from those soils have shown increasing irregularities in health and reproduction; our observations are pointing to correlations among the above factors; soils, crops and livestock.

NITROGEN FOR PROTEINS
AND PROTECTION AGAINST DISEASE

Provisions of our food and feed proteins is a serious problem. Getting meat, milk, eggs, seeds and similar body-building foods has long been a major part of the struggle to feed ourselves. An adequate supply of protein supplements is the problem in feeding our animals. We are not satisfied to nourish ourselves with seed proteins only. We want and need proteins of animal origin too. Not even all the animals can get along on only vegetable proteins as chickens and hogs show. Grains contain protein in their germ, and meat, milk solids and eggs are made up almost wholly of that essential food constituent, yet we have given little or no thought to the supply of nitrogen in the soil required for the plant's fabrication of protein and the solution of the food problem. We have not thought of nitrogen in the soil as necessary for the synthesis of different proteins. Much less have we given proper credit to proteins for the multilateral protection they provide against hunger in the life struggles of microbes, plants, animals and man. How to accurately measure proteins is still an unsolved problem, even though nitrogen is accepted as synonymous with proteins, the measurement of nitrogen alone has not been a satisfactory gauge of the quantity or quality of a protein. This is suggested by the fact that nitrogen is generally regarded as constituting about 16% of protein, since in determining the protein content of a substance we usually multiply its nitrogen content by 6.25. The cereal chemist multiplies the nitrogen by 5.73 because he says that cereal protein has over 17% nitrogen.

This indicates that nitrogen is converted into organic compounds of carbon, hydrogen and oxygen to make proteins with different ratios of carbon, for example, to nitrogen. The very components of proteins are therefore highly variable. Consequently the process of igniting a substance in sulfuric acid, measuring the nitrogen obtained and multiplying it by a single arithmetical factor like 6.25, is not a very accurate measure of the protein. If it is the nitrogen that is significant in relation to the other constituents of the protein molecule, then surely the total nitrogen, including other than protein forms, handled by a mathematical factor in common usage by a majority of analytical chemists is not an accurate gauge of protein. Measuring this essential part of our foods is still a problem, with much to be learned for our better nutrition and improved health.

How proteins are put together from the elements is another baffling problem. This consists of combinations of smaller units, or amino acids, which, although they are called acids and will react with alkalies, are also reactors with acids. In that sense they are of dual nature as to chemical reaction. Much has been learned about the chemical structure of protein, but how nature starts with the separate elements, i.e., carbon, hydrogen, oxygen, and nitrogen, and builds an amino acid is still unknown. How sulfur and phosphorus are put into the protein molecule is another unsolved mystery of the creative processes of growth. While we can separate a protein into its amino acids, and can separate them into their chemical elements, this gives no clue to the processes by which they are put together in nature. We have to depend upon nature to guide the chemical synthesis of them. When we speak of animal proteins and vegetable proteins we do not mean to suggest that the animal has built them from the separate elements. Animals only assemble the proteins. They must find the required amino acids for that purpose in the vegetable or plant proteins they consume. Apparently it is their inability to create proteins that makes higher animals dependent on plants and microbes supposedly below them in the evolutionary scale.

Microbes and plants can take the elements, or simple compounds of them, and build them into amino acids. Although in simplest terms it appears that the plant uses air and water, under solar power, to make carbohydrates and then uses nitrogen from the soil or from the air to make amino acids, the process is far more complex. Were it not so, proteins might not be a problem. Instead they should be as plentiful as carbohydrates. All plants are made up mainly of carbohydrates. Microbes and plants are able to synthesize proteins from the elements, but not without a struggle, for it is a problem for them too, even to make the simpler proteins by which they live but which alone would not support us or our animals.

Microbes synthesize only limited kinds of amino acids and limited combinations of them as proteins. Some protein products may be very important to us but microbes require more than their component elements for their fabrication. Apparently more is required for protein elaboration by plants than the combination of nitrogen, sulfur and phosphorus from the soil with carbohydrates. Legumes, that are particularly appreciated as synthesizers of protein, require calcium, phosphorus, potassium, magnesium, boron, manganese, copper and other elements from the soil for their growth processes. When we say that these better forage feed crops are "hard to grow," we mean that it is difficult to get these plants to produce proteins. These essential food constituents for us are not readily manufactured. They require not only nitrogen (so commonly deficient) in the soil, but they also require soils fertile in more respects than is commonly recognized. It is natural to expect, then, that microbes with simpler requirements for growth will produce only the simpler proteins. Plants, which we think of as easy to grow or which are growing on less fertile soil, will also create the simpler proteins, proteins simpler than those created by plants failing to grow on such soils. Shall we not then expect a larger variety of amino acids and a greater total of them, in plants growing on soil better supplied not only in the major nutrient elements, but also in what are commonly thought of as trace elements?

If animals have more complex food requirements, particularly for the amino acids of proteins which they cannot synthesize, it is not difficult to believe that an extensive list of soil fertility elements will be required to synthesize the various kinds of proteins they need. Man doubtless is in the same category as the animals. If the soil, then, fails to provide adequate supplies of the required fertility elements, shall we not look for deficiencies first in the variety of proteins needed by animals, and second in those proteins needed to grow the plants commonly considered as being more nutritious as animal feeds? If so, is it not reasonable to suggest that the output of proteins in terms of the supply of the different amino acids and of the sum total as protein is a direct reflection of the fertility of the soil? If animals are to obtain all of the amino acids they need, they must gather their feeds from many simple sources and many soils, or from a single soil that is highly complex in its content of fertility elements. Complete proteins, therefore, in terms of animal and human nutrition, require a very fertile soil. Conversely, it is not surprising then if a shortage of nitrogen, phosphorus, calcium, boron, manganese or of any other essential nutrient element in the soil should produce crops in which the total protein content though high, might be inadequate in respect of the required amino acids needed by animals or man for good substance and good health.

The production of proteins, in their fullest amounts and variety through plant growth may constitute a complex demand upon the fertility of the soil. Nevertheless, there seem to be no greater values to be had from soil treatments than the contributions they make to the elaboration of proteins in the crop. Traditionally, nitrogen has been used to grow larger crops, but should we not give some consideration to its undoubted association with the protein content in the crop? Nitrogen and proteins are synonymous in the mind of the chemist, but nitrogen put into the soil has not been a guarantee of the fullest amount of protein or of the food quality of the subsequent crop. It has not been related to the complete array of amino acids to provide a balanced protein for animals and man. We may well give attention to the services and functions which proteins render in growing the body, whether that body be microbe, plant, animal, or man. We need to give more attention to proteins for their services in keeping bodies in good health by protecting them against the invading forces of disease. Carbohydrates build plant bulk, but neither they nor the fats carry the power to grow. Only the proteins can propagate themselves, transmit life, multiply themselves and regenerate new cells by their own division. Life chemistry is carried on by means of the proteins. Some we call enzymes. They give speed to chemical reactions. Some we call hormones. They coordinate our body activities. Some we call viruses, producers of diseases. And some we call antigens that serve to protect us against disease. But all of them are of such chemical composition as to be classified as proteins. "Good healthy growth" in young people, we say can be had only by their consumption of plenty of proteins. Milk is the food commonly used to supply them. Recovery from sickness calls for protein rich food. Tuberculosis is now arrested, and "cured" by a high-protein diet and rest. Protein in nutrition has come to be protection and guarantor of human good health, but we have not yet been ready to believe that nitrogen in the soil along with other fertility can similarly protect other forms of life like plants rich in protein by which they protect themselves. That fertile soils make plants rich in protein by which they protect themselves against fungus diseases and insects is a fact not yet accepted even though we are beginning to accept the idea that protein is protection for mankind.

Increasing the calcium content of a clay-sand medium growing soybeans demonstrated their increasing freedom from an attack by a fungus resembling "damping off." This increase in calcium in the medium was brought about by merely increasing the clay content in the sand. The clay was one of standardized proportions of exchangeable calcium and hydrogen leaving it acid at a pH of 4.4. Merely increasing the acid clay, and thereby the available calcium through root contact with more clay and calcium for larger amounts of this in the crop and for more nodulation and nitrogen fixation in the plants, was the sole difference between complete immunity from fungus attack and the complete destruction of soybean plants.

In another demonstration, more protein in a spinach crop increased its ability to protect itself against the attack of leaf-eating thrips. Here the nitrogen offered per spinach plant was varied through a series of 5, 10, 20 and 40 millequivalents (ME). Each of these levels of nitrogen was also combined with a series of 5, 10, 20 and 40 ME of calcium, the element commonly associated with the elaboration of nitrogen into the protein compounds synthesized by plants.

While all the plants were equally exposed to the thrips on the weeds growing nearby, the insects attacked only those spinach plants given the lower amounts, namely, 5 and 10 ME of nitrogen. Even on these, the attacks were less damaging as the amount of calcium associated with each unit of nitrogen was increased. With replications in which the content of calcium and nitrogen was increased ten times there was a clear cut demonstration that as these two factors, nitrogen and calcium in the soil were increased to favor increased protein synthesis by the spinach plants, there was increased protection to the point of immunity against the insect attack. Here more protein within the plant provided more protection not against "disease" as it is commonly considered but against an invasion of and consumption by insects. Here was the suggestion that nitrogen, calcium, and other chemical fertility elements put into the plant meant not only protection but also higher food values. This constituted a far more effective escape from the insects than could have resulted from complex pesticides which offered no nutritional values to the plants.

One may well raise the question whether an increased supply of available nitrogen and of other fertility, including the trace elements, in the soil is protection for crops because of the production of more of any one kind of amino acid and simple protein in the plants or because of the production of a more complete array of the different amino acids making up the proteins. When some trace elements seem to be more effective for animals and perhaps humans in consequence of their intestinal microbial synthesis into compounds; when trace elements in soil seem to come through the corn grain to encourage apparently healthier liver tissues in test rabbits; and when these same elements increase synthesis by alfalfa of the amino acids commonly deficient in corn, it appears to be evident that something more is necessary to provide a complete explanation than an arithmetical formula. Nitrogen can be synonymous with protein and protein can mean much more in the way of protection against disease— even against insects—when once we understand more completely what a truly fertile soil is.

DECLINING NITRATE LEVELS IN PUTNAM SILT LOAM

A 13 year study of the nitrate content of the surface soil of Putnam silt loam under cultivation reveals a gradual decline in the levels of the supply of this plant nutrient. The soil in question is a rolling phase of Putnam silt loam in a transition area approaching Lindley loam. It had been in grass for many years as pasture and as an undisturbed part of the station field previous to its use in this project. It was plowed as probably a native sod before this study was undertaken.

During 13 years, 1920-32, inclusive, bi-weekly nitrate determinations (except when the soil was frozen) were made on the surface soil of a series of small plats. Some were continuously in corn, some in wheat, and some in fallow. Different tillage and fertilizer practices were combined with these crop treatments. During the first six years the wheat plats were plowed at different dates. The corn plats were given different tillage, while some of the fallow plats were given straw mulches. No plant nutrient additions were made. During the last seven years all the plats were given standard tillage treatments supplemented with the more common fertilizer applications. As an illustration of the general behavior of these nitrate levels, three plats have been selected as representative. One of these was in corn continuously for 13 years with the normal tillage treatment of spring plowing and surface cultivation, but given a spring treatment of ammonium sulfate (25 pounds per acre) broadcast during the last seven years. Another plat was in wheat continuously during the same time with regular plowing on August 15, followed by the customary treatment of seedbed preparation. During the last seven years this was fertilized with the equivalent of 200 pounds of 2-12-0 at seeding. The third plat was fallow with a six ton straw mulch applied in late April following the spring plowing and after the removal of the straw in time to permit drying of the soil to a tillable condition. During the last seven years, the equivalent of a two ton application of green manure in the form of chopped sweet clover of alfalfa was turned under each spring.

The numerous nitrate determinations have been assembled as monthly averages per advancing five years. Only the data for the growing season, March to October, inclusive, are presented. They are given as graphs in figures 1, 2, and 3 for these three respective plats for four of the nine advancing five year averages. The nitrate levels manifest certain cycles of seasonal variation with maxima at certain seasons and minima at others, each characteristic of the crop and the time of its seedbed preparation. For corn, the maxima are in the late spring and the minima are in the fall; while for wheat these occur in early spring and early summer, respectively. The date of the season's maximum for corn varied, although the minimum occurred regularly in September. Under wheat, the season's maximum was found regularly in the early spring, while the low point occurred in June or July. The range between these two points was about the same for these two crops in the same year, but became less with succeeding years. The minimum under wheat was lower than that under corn. Under the fallow soil covered with the straw mulch, the nitrate levels were low at the beginning of the study in comparison with those under wheat, because of the depressive effects of the straw mulch. The seasonal cycle consisted of a small rise in May in consequence of the preceding removal of the straw mulch, the drying of the soil, and its tillage by the mould board plow. Following this rise, there was a gradual decline to a minimum in late summer.

Perhaps the most significant fact established by these data is the decline of

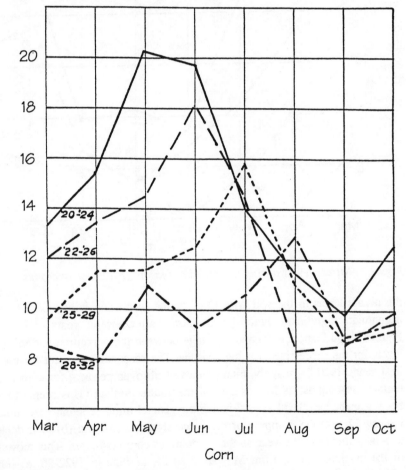

Figure 1. Nitrate nitrogen in the soil in corn. (Five year averages during the growing season.)

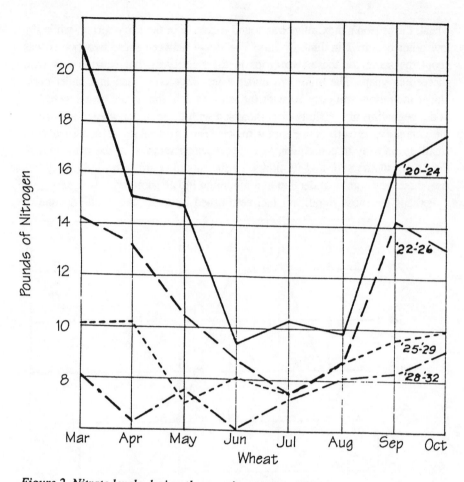

Figure 2. Nitrate levels during the growing season. (Five year averages.)

the nitrate levels in the soil with time, or the increasing failure of the soil to accumulate significant quantities of nitrate after several years of cropping. Figures 1 and 2 show a decreasing range between the maximum and minimum levels for the successive growing seasons. The seasonal low point for the earliest years, 1920-24, namely, nine pounds of nitrogen per acre for corn in September, was almost as low as any of the future minima. This seems to have been near the base level toward which the general nitrate content was falling as indicated by (a) the dropping of the successive maxima, and (b) their delaying to a later month in the year as the cropping to corn continued. This movement of the maximum point from May in 1920-24, to June in 1922-26, to July in 1925-29, and to August in 1928-32 suggests that the nature of the organic matter source of nitrate nitrogen was changing, possibly to those forms demanding

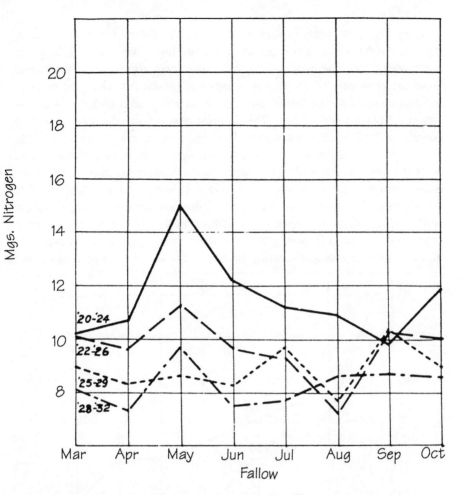

Figure 3. Nitrate levels during the growing season. (Five year averages.)

higher temperature for significant nitrate release.

Under wheat a minimum point was also suggested at the early part of the study, though it did not so clearly indicate itself as a base level, as was true for corn. Later minima under this crop fell below that found at the outset. It is interesting to note that under wheat these lowest levels were significantly below those of corn. Even under the fallow soil (figure 3) the maximum levels of nitrate became lower with time.

The decline in nitrate levels in the soil of these three plats can be measured more effectively by using the average figure of pounds of nitrate nitrogen per month during five successive years and averaging the eight months for the growing season of March to October, inclusive. Such growing season and monthly averages per advancing five year periods are exhibited in graphic form in figure four. It is clearly evident that the nitrate levels fell faster with increasing years under wheat than under corn. This decline is very marked, however, under both crops. The nitrate levels under these two crops, starting together at about the equivalent of 15 pounds of nitrogen for the first five year monthly average, dropped to the final average of about ten pounds under corn and to almost 7.5 pounds under wheat. The corn plat represented a decrease of about 5 pounds, or 33%, while the corresponding figure under wheat was 7.5 pounds, or 50%. On the fallow soil, the nitrate level was much lower at the outset on account of the disturbed nitrogen-carbon ratio following the application of the straw. Even with the early low nitrate, this soil shows a falling nitrate level in the absence of a crop. This decline is marked during the latter part of the period, which occurred in spite of the addition of the equivalent of two tons of green manure annually during the last seven years. This addition seems to manifest some effects as shown by an irregularity in the form of a rise in the

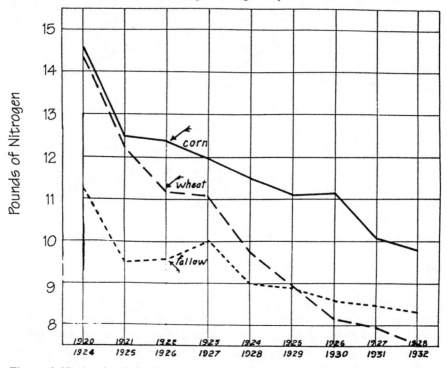

Figure 4. Nitrate levels by five year, growing season, monthly averages.

curve for the two 5-year periods of 1922-26 and 1923-27, which include the first years after 1926 when this green manure treatment was begun.

   At the close of the nitrate study the plats were seeded to a mixture of red and sweet clovers drilled with light applications of limestone combined with fertilizer treatments. The first year stands (October 1933) of these legumes on the different plats reflect the declines in fertility represented by the nitrate levels according to the stand counts of the clovers on these plats as areas 3.5 feet square. On the several wheat plats the stand count average is 40 plants, on the corn plats 55, and on the fallow plats 51. These figures do not show the magnitude of the differences since they do not reveal the size of the plants. The total dry weights of the clover plants harvested from eight separate square foot areas on each of the particular plats reported herewith were as follows: On the corn plat 9.0 grams, on the wheat plat 2.9 grams, and on the fallow plat 28.8 grams.

These data emphasize with significant force the lowering of the nitrate level with time and at rates varying with different crops. Further, they point out that this decline is little disturbed by common fertilizer applications and continues under fallow even in the face of relatively heavy applications of green manure. The data suggest that much nitrate production is going on at the expense of the virgin organic matter in the soil. Such changes in fertility levels reflect themselves in variable clover stands and in the crops following, suggesting that other fertility factors besides nitrogen may be following a similar decline and may be associated with the causes in operation.

FERTILIZING WITH NITROGEN:
THE COW MAKES HER SUGGESTIONS

Among the garden crops, beans, lentils and other legumes are mainstays because, as nitrogen-fixers, they have long been known to be higher in the proteins and all else which such plants must take from the soil as essential elements and compounds to be legumes. These are the crops which can take the gaseous nitrogen from the air and combine it into the highly valuable food constituent, namely, the proteins. In feeding animals, the legumes are the feeds for "growing." They supplement the corn, for example, as feed for fattening the animals—usually castrated males carrying no load of reproduction or of growing fetal young. Because proteins contain the element nitrogen in a particular chemical structure which we call "amino" nitrogen, and some other forms of it, to the extent of about 16% of these compounds, we have been measuring the nitrogen by "ashing" our foods and feeds in sulfuric acid. Then we multiply the measured value of this element by 6.25 and call the result the "crude" protein of the food or feed. The cereal chemist multiplies it by 5.73 for the proteins in the grains. Thus, the nitrogen has become synonymous for the proteins, but not without danger of serious confusion.

That confusion is becoming more dangerous, now that we are applying nitrogen to the soil generously as fertilizer, and such in highly active chemical forms. The plant may take up these; it may fail to convert them into the proteins; and yet our chemical analysis will credit the plant with high protein while much of that may even be poisonous. There is also the increasing inference that non-legume crops, fertilized generously with nitrogen and consequently brought to contain a high percentage of nitrogen in their dry matter, must therefore, be rich in proteins. If that figure for percentage of nitrogen duplicates the one for legumes, the erroneous conclusion is apt to be drawn that our use of plenty of fertilizer nitrogen on grasses and other non-legume crops will grow proteins in quantity and quality in those as well as when we use the regular legumes.

That such is not necessarily the case follows from the fact that soils must be well stocked with calcium, magnesium, potassium, boron, manganese, copper, zinc, and many other essentials in the list of the major and the minor or "trace" amounts before legumes will grow. Only after legumes have a balanced soil fertility in terms of all the essential elements except nitrogen, and those in good supply first, will those plants add the nitrogen of the atmosphere to that stock and carry out this process of nitrogen fixation by which they grow, protect themselves, and reproduce. The nitrogen is not taken up in that natural process by legumes except as it moves via the protein forms of it. That the corn plant, as one of the grasses, can deceive us and take up large quantities of nitrogen from the soil without converting this essential element into protein, was demonstrated by the drought when the so-called "silo-filler disease" gave us not only a newly coined term but also some new concepts about the dangerous biochemistry of silage making. The chopped fodder gave off enough nitrous oxide fumes over night to kill one on entering the partially closed silo the following morning. The chemical analysis of the cornstalks suggested that, because the heat of drought had destroyed the enzymes which were normally converting nitrates coming into the plant from the soil into other plant compounds, the nitrate nitrogen or even this in the nitrite combination had accumulated excessively in the corn stalks.

Nitrates and nitrites are deadly compounds for animals and man when taken in even very small dosages. Some analyses of corn stalks, reporting 1% of total nitrogen, showed as much as 0.65% of this element in the nitrate form. Thus, we might have believed that the dry matter of the corn fodder had 6.25% protein. But instead it had 4.06% so-called protein—equivalent of the 0.65% nitrogen—which was poison, as the deaths of the cows had sadly reported. The cow has, perhaps, been the oldest producer and distributor of nitrogenous fertilizers, putting out some of it in the organic and some in the soluble salt form, viz. urea. She demonstrates her activities in that industry every time she voids her droppings of either feces or urine on the grassy swards. There we see the markedly green growth of it encouraged by the urine.

But if we observe closely and consider farther, we will be forced to conclude that such is an excess of nitrogen fertilization in the humid soil areas, and is not producing a nutritious forage feed, regardless of how demonstrative in yield and how attractive to the eye. Testimony to that is given by the cow herself when she refuses to eat that green growth, but lets it grow taller while she eats the surrounding grass shorter and shorter. Even she tells us that one must be cautious to balance the nitrogen salts with all other fertility in using them as fertilizers. The cow is not classifying forage crops by variety name, nor by tonnage yield per acre, nor by luscious green growth. Instead, she is classifying forage according to its nutritional value in terms of complete protein and all else coming with it, according as the fertility of the soil determines. As a biochemist or an assayer, she is not satisfied with the value of the nitrogen in the ash multiplied by 6.25 and labelled "crude protein."

When we fertilize with nitrogen salts, we dare not always say we are thereby necessarily making crops more concentrated in protein because they look greener and more luscious. The growing of plants is a problem in plant nutrition no less complicated than any other problems of trying to feed some life form properly. So when using nitrogen-carrying fertilizers, one dares not operate under the belief that "if a little is good, more will be better," even though nature seems to follow that philosophy when she fertilizes with nitrogen in the form of organic matter which she may pile up abundantly.

FERTILIZING WITH NITROGEN: RABBITS TESTIFY BY EXPERIMENTS

When we observe crops either in the garden or in the field, they are anatomically very similar. They all have roots, stems, branches, and leaves. They are of a luscious green color. They look alike, and—so far as we can see—we might expect one to do what the other does when we fertilize the soil under them with extra chemical nitrogen. We are usually delighted when both the garden and the field crops improve in their production of more vegetation and a greener color. We measure their contents of nitrogen and find that the concentration of this in the dry matter was increased decidedly by the fertilization. We believe we have thereby made them richer in protein as food. We do not appreciate the fact that, in terms of their physiology, the plants may be as widely different as the talents of folks when some can and some cannot. Two plants looking alike are not necessarily of equal feed value. The cow confirms this when she may choose one plant in her grazing and disregard the other. In the form of hays separated in the feed-rack, she may consume one entirely before even touching the other.

Total nitrogen and essential amino acids in timothy and red clover hays (mg/g per gm. dry matter)

	HAYS	
	Timothy	Red Clover
Total nitrogen	13.10	17.40
Methionine	0.61	1.31
Tryptophane	3.11	*
Lysine	1.66	5.26
Threonine	2.97	4.51
Valine	3.49	5.70
Leucine	14.00	6.07
Isoleucine	7.12	4.43
Histidine	0.81	1.37
Arginine	3.64	9.59
Phenylalanine	2.79	4.86

* Not measured.

Under most delicate testing by using weanling rabbits in feeding experiments, the fertilizing of a grass like timothy with nitrogen increased the "crude" protein decidedly. But even then, this was not necessarily good nutrition. This suggests that the physiology of the non-legume, timothy, in its use of fertilizer nitrogen, does not necessarily give us the nutritional values for rabbits in the crude protein or even in the amino acids which are equal to those in a legume like the red clover.

Timothy hay fertilized with both nitrogen and trace elements was fed experimentally to weanling rabbits during the summer heat wave of 1954. But with every severe rise in temperature the heat killed some of the rabbits until 70% were dead. Then, since the experiment had gone its planned time and the rabbits added from the stock supply at fortnightly intervals had maintained the number, dried skim milk powder was added as a protein supplement to improve the quality of the ration in that respect. From that date forward no more fatalities occurred, even though the heat waves continued. Then the original experiment was repeated. There resulted again the repetition of the deaths by the heat waves, which were accepted as part of the test until 30% of the experimental rabbits were killed. Then the timothy hay was replaced by some red clover hay grown in a field where the clover has been in rotation systems since 1888. From the date of this substitution, and with the heat waves holding on, no more rabbits died while feeding on the red clover hay.

It is interesting to note that the timothy and the red clover hay, under chemical analyses for total nitrogen and under microbial bioassay for the essential amino acids, would not tell us all the reasons why the clover saved the rabbits from death by the heat but the timothy did not. It is interesting, nevertheless, to note the differences between the two hays so far as the chemical analyses for these essentials could give suggestions. They are given in the accompanying table. From such data one cannot see differences wide enough to make a guess as to the particular amino acid deficiency which killed or the particular sufficiency which saved. It is significant to note how similar the two hays were in totals of the ten and the nine amino acids respectively, but yet were more widely different in the total nitrogen. In this latter respect the red clover was almost a third higher than the timothy. As an additional experiment, the red clover hay grown on 11 plots with that many different soil treatments and history, were each made up into diets for test rabbits so that each of the series contained 1.31% total nitrogen, or 8.18% crude proteins supplied by the clover from the particular plot. All else, as dietary factors, were also brought to a constant as nearly as possible. Then this set of supposedly uniform diets was again fed to carefully selected weanling rabbits and their gains in weights taken as the index of the diets' efficiencies, even when all were the same in "crude" protein.

FERTILIZING WITH NITROGEN: WE MAY USE TOO MUCH SALT

Animal manures were the first soil treatments to give improved growth of the corps. The accumulated manures of sea birds collected and imported from the arid Pacific Coast of northern South America were some of the first commercial fertilizers. They were sold to our cotton farmers under the trade name of "guano." From near that same source and representing, probably, the evaporated leachings from the guano, there came later the nitrate of soda or the Chile saltpeter. Thus, history gave us a gradual shift from the fertilizer nitrogen in the organic manures to that in the form of the inorganic chemical salts. But this shift was not accepted without serious protests by the southern farmers. They contended that guano gave better results on their crops than the saltpeter. Perhaps, even they recognized some fertilizing helps from the organic compounds in the guano which were not given by the simpler, purer, more concentrated chemical salts. Many are still contending that organic manures are better fertilizers than pure chemical salts.

Not too long after the sodium nitrate became the common nitrogen fertilizer, the ammonium form of nitrogen combined as the sulfate came into the fertilizer trade. This was collected, along with coal-tar wastes, as a byproduct of the destructive distillation of coal to give coke for the reduction of iron ores. This was a form of nitrogen different from nitrate. Chemically, it was a positively charged ion. Consequently it is held in the soil against rapid loss to the rain water going through it. It undergoes a slow, microbial conversion there into the nitrate form which has the negative charge, and when united with a positive ion, like calcium, for example, is not so firmly held by the soil. It is then a highly active salt and may be readily washed out. It is taking other ions of nutrient value and of opposite charge, as well as calcium, out of the soil. While this all happens, the nitrogen shifts itself from the positively charged ammonium form, held by the soil, into the negatively charged nitrate form not held. If it is applied as a sulfate, this latter, as a negative ion, also carries calcium or other positively charged ions out. The ammonium ion as a positive one tastes hot. The nitrate, as a negative one, tastes cold. They are opposites in more ways than one.

When these various salts of nitrogen were combined with superphosphate as mixed fertilizers, it was soon discovered that while one could make heavy applications of phosphates with the seedling of grains, the nitrogen fertilizers could not be so used without injury to the germinating seed and serious reduction in the stand of the crop. Then when a potassium salt, like the muriate, was also included in the fertilizer salt mixture to give the so-called "complete" fertilizer, the damage was still worse. On Sanborn Field one plot started as early as 1888 with heavy applications of commercial fertilizer salts of nitrogen phosphorus and potassium, enough to represent those three elements taken off by the grain and straw in a 40-bushel crop of wheat. Beginning with the virgin soil, that much of chemical salts drilled with the wheat seeding, where this crop was grown annually, was not damaging to either the germination or the final stand until after nearly 20 years. But during that period with all the straw and grain removed and with no organic matter returned, the soil has lost much of its humus and organic matter, which represents its capacity to "buffer" the shock of so much chemical salts. Even after that period and even today, the total fertilizers can be applied safely only as a divided application, one part with the autumn seeding while the other parts must be applied in the spring or later.

In practice, in general, most fertilizers are now applied on soils unable to handle these more concentrated forms and heavier salt treatments, unless by special placement of them at some distance from the seeds. We are faced with the mechanical problem of proper placement of those chemical salts to escape their early damage. That was no problem with nitrogen in guano or in its organic combination. Such facts are turning our thinking back to organic matter to consider the problem of handling nitrogen and potassium, or other salts, so highly active and thereby so readily damaging. We are taking to the suggestion that, perhaps, a kind of natural composting of the nitrogen fertilizers by the help of the microbe in our virgin soils of high organic matter contents was nature's method that saved us when we first used it without damage from the excessive salts.

There comes, then, the suggestion that nitrogen ought to be used along with crop residues, and with applied organic matter if the depleted soils are to be restored and the nitrogen held to the best advantage in the soil. Composting either within the soil, or in the special compost pile above it, seems to be the prevention against fertilizing with too much salt. This seems to be no small danger when the concentrations of the fertilizers are mounting and the larger applications commonly applied are both crowding under increased economic demands for higher yields per acre. It is these conditions which portend a compulsory reverse of history with its shift from the nitrogen salts back to more of these elements in the form of organic compounds.

FERTILIZING WITH NITROGEN:
FERTILITY IMBALANCE AND INSECT DAMAGE

Just as wild animals seem to know how to protect themselves and to be healthy, so plants, in the wild, seem to grow their own protection even for the seed carried over the winter. That the soil fertility should help grow a protection into the seed against grain insects is scarcely surmised until we see it demonstrated. During the season of 1954 some trials with high nitrogen fertilizers on corn in South Dakota furnished the samples for some careful observations of the failing protection against the lesser grain borer when the corn was fertilized with nitrogen only. In the illustration on page 116, the small ear, the middle one of the three, represents the size of the ears and suggests the yield of the corn crop on the soil given nitrogenous fertilizer only. This was the treatment suggested by the soil test which indicated no serious fertility deficiency save that the nitrogen might be improved by soil treatment with this nutrient element as fertilizer. The nitrogen was applied with the corn planting in the spring of 1954. The grain was harvested in September. The ear on the left represents the yield of corn from the plot given, not only nitrogen at the rate used on the other plot, but also this combined with fertilization by phosphates, even though the soil test had not suggested that there was a deficiency in the phosphorus. Both of these grains were grown from hybrid seeds. The samples from these two soil treatments had been stored for two years when the photograph was made in late September, 1956, to illustrate the wide difference in the damages to the seeds by the lesser corn borer. The ear grown on the soil fertilized with nitrogen only was almost completely riddled with accumulated, floury waste removed twice in advance of the last accumulations exhibited in the illustration. The ear on the left, grown on soil fertilized with phosphates as well as with nitrogen, had been in contact with the badly infested one but was damaged by the grain borer only at the point of contact with the smaller ear, and on the lower left side at a few points of contact, apparently, with the shelf. The ear on the right was grown on well fertilized, highly organic soil from open pollinated seed and had been in contact with the badly damaged ear for six fall and winter months before all were photographed. It showed but one or two grains attacked by the borer.

It has been previously demonstrated in experimental trials that young soybean plants protected themselves against a fungus attack (suggesting "damping off") as the increasing amount of clay in the soil supplied them with increasing amounts of exchangeable calcium while all other fertility supplied were constant. See page 101. It has also been demonstrated that the attacks of the leaf eating thrips insects were prohibited when the nitrogen applied to the soil was in larger amounts, and when the lower amounts of that fertilizer allowed the insect attacks, even those were lessened by increasing amounts of calcium used as fertilizer. This demonstrated also that the fertilizer treatment of nitrogen was better protection against leaf eating insects according as the application was heavier and in combination with calcium. In the case of the stored grain, there was the better protection against the grain borer according as the nitrogen—synonymous with protein—was balanced by some phosphorus, but was inviting destruction almost completely when it was used alone as the fertilizer. It is significant to remind ourselves that when forage crops use extra nitrogen from the air to build themselves richer in protein, which is the protection or immunity in most living forms, they do so only after all other essentials of fertility are present in the soil in generous amounts. In the humid area "legumes are hard to grow," or the soils must be well balanced in their fertility if these crops are to fix nitrogen and are to be free of diseases. We dare not rush in to apply nitrogen, then, on the non-legumes—if this reasoning by simile is sound—without anticipating the possible imbalance to be damaging rather than beneficial. If nature fertilizes with nitrogen via the plants' combination of it in the form of organic matter, perhaps the composting of fertilizer nitrogen will be wiser use than its direct application to a crop as salts. We may well ponder that question in relation to what nutritional and protective values the crop finds in the nitrogen treatments as generously applied salts.

2

WATER AND SOIL FERTILITY

WATER—MAJOR MINERAL OF SOIL NUTRITION

We have not commonly considered water as the major mineral of the earth's surface. In its service to man and animals we have not listed it as the one mineral among the others, which breaks them down physically and chemically, and takes them to itself in suspension and solution. This is done first on land and then in the sea. Within the soil and among the three inorganic soil separates of sand, silt and clay, water may well be viewed as the fourth soil separate. It is the most finely divided one, even of molecular and ionic dimensions.

It is in its three forms of solid, liquid and gas (all possible within the temperature range of 100 degrees Centigrade), that water is unique. It is the most active mineral in making the other minerals—including rocks as combinations of them—dynamic in the soil while they are en route to the sea. It is true, then, that water is the major mineral of soil.

Since water is a liquid mineral at ordinary comfortable temperatures; it is the medium within and by which the many solid mineral compounds are decomposed. The action of water sets their inorganic elements free. Water puts them on their own activity for chemical combinations with others, both organic and inorganic. The magnetic minerals of the earth originally contained no nitrogen and no carbon. They had none of these two elements by which the organic, or living matter, is distinguished from the inorganic, or dead. These two organic requisites had to be taken from the atmosphere initially to be put into the soil. Water serves in reducing rocks to sand, silt, and clay, to the dust into which the moist breath of creation, containing carbon and nitrogen, is blown when that process occurs.

Regions of rainfall, ample in both total per year and favorable distribution through the warmer, growing season, are in that favorable category because water serves to hasten sufficient rock decomposition to give the change of reserve minerals into colloidal clay; to give soluble elements at the same time which may be adsorbed on and held by the clay-organo-colloid. It also passes these elements by exchange to the plant roots for their acidity or hydrogen in the enshrouding carbonic acid from their respiratory carbon waste.

More than two-thirds of our crop production still comes from this development of soil more than it comes from the chemical fertilizers we apply for the bigger crop yields. Water in any climate favorable for agricultural production is making it favorable more because the water is serving as a chemical reagent than it is in providing crops with drink and cooling waters in the plant's stream of transpiration.

When we think of water taken from the soil by crops, we are prone to emphasize its movement from the soil into the plants and its evaporation, or transpiration, from the surface of the leaves. Also, so much is said about needing the "available" inorganic nutrient elements from the soil, one is often led to believe that these essentials are all in solution in the soil water. So, since water moves from the soil into the roots and up to the leaves to be evaporated there, we are apt to infer that the flow of water must occur to bring the fertility elements from the soil into the plant and to deposit them through evaporation of the soil.

The facts are quite the opposite. Nutrient elements move from the soil into the plant root because of, first, their concentrations and ionic activities in the moisture film on the soil, and second, because of the nature of the cell wall of the root hair, coupled with the metabolic behavior of the contents of that cell and the many others connected with it.

These migrations of the fertility essentials from the soil into the plant are governed by the soil's store of them and the plant's physiological processes of moving them from the root hair into other plant parts. Evaporation of water at the leaf surface is not the cause of movement of "available" nutrients from the soil into the crop. The flows of plant water and of nutrients are in separate streams, independent of each other.

That the water of transpiration flows independently of the nutrients maybe easily demonstrated. A potted plant in the sunlight under a sealed glass bell jar, containing carbon dioxide in an atmosphere saturated with water vapor, will grow nicely by taking water by transpiration. Plants on infertile soils, especially those deficient in calcium, will transpire water from the soil and grow with an increase in bulk, but will give nutrient movement back from the planted seed supply to the soil.

In demonstrating the first of these illustrations, the transpiration stream was stopped, but yet the stream of nutrients flowed on from the soil into the plants to keep them growing. In the second illustration the transpiration stream flowed on as usual, but the nutrient stream reversed its course and moved in the direction opposite that of the flow of water. These are facts enough to prove that the transpiration of water from the leaves does not carry the nutrient elements from the soil into the plant.

Transpiration of water obeys the conditions of the atmosphere inviting the evaporation against the forces within the soil. Such evaporation serves as a kind of air conditioner to hold down the plant's temperature. It also maintains the moisture in the plant as a medium in which the inorganic elements may be ionized and chemically active. But the flow of water through the plant in the transpiration stream of this liquid mineral is not the cause of flow of the fertility elements into the plant. The flow of these for plant nutrition is quite independent of that of water going in the same direction. Here water as a mineral does not—by its physical behavior as flow—influence the other minerals to behave likewise. But by its presence as the medium it sets the stage for the elements separated from their mineral compounds to be active in plant nutrition.

When we define soil as a mixture of inorganic matter resulting from the disintegration and decomposition of rocks and their component minerals, we fail to include water as a major mineral. Perhaps its liquid nature and its quiet behavior make its services so unnoticed that we take its presence for granted. However, as the pressure of population for production of more crops increases as the regularly stored supply of water in the soil decreases,we shall gradually recognize water as the major mineral of the soil for our nutrition and survival.

LESS WATER REQUIRED WITH ADEQUATE FERTILITY

In addition to sharply increasing corn yields, full soil treatments were an important factor in materially reducing the amount of water required to produce a bushel of corn in experimental trials this past summer [1953] near McCredie in central Missouri.

Dwight D. Smith, research associate in the University of Missouri department of soils, reports corn on soil with full treatment—grown in a four year rotation of corn, wheat, and two years of meadow, with grain and hay crops removed—produced 79 bushels per acre and required 16 inches of soil moisture. In terms of water use, some 5,600 gallons of water were required to produce a bushel of corn.

Corn grown without soil treatment—in a corn-oats rotation with only the grain removed—produced 18 bushels per acre and required about 14 inches of soil water to raise the crop. On a per bushel basis, some 21,000 gallons of water were required.

Measurements of the use of soil moisture from the upper 3.5 feet of soil were made. Root development under fertilized crops was greater and they penetrated deeper into the soil. Thus, corn grown on plots with full treatment had root development sufficient to take water at a greater rate. And since the roots penetrated to greater depth, the plants had a greater supply available from which to draw water than the shallow rooted corn on the unfertilized soil. By early August, plant roots in the fertilized area had penetrated below the 3.5 feet depth. So these plants actually took a little more water from the soil than could be measured in this experiment.

Water removal from the unfertilized plot was about one-eighth inch daily throughout the season. This was also the approximate water removal from the full fertilizer treatment plots for the first one-third of the season. After that time, the removal was more than one-quarter inch daily. Evaporation of moisture from the soil was an important factor during the early growth stages. As the corn grew the crop required more water, but shading of the soil cut evaporation losses.

The 1953 yield differences were quite similar to those of previous years. During the past six years, average yields in the corn-wheat-two years meadow rotation have been 100 bushels of corn, 23 bushels of wheat, and a little more than two tons of hay per acre.

In the two year rotation of corn and oats without fertilizer, average yields have been 23 bushels of corn, and 6 bushels of oats per acre. However, with a similar two year rotation of corn and oats, but on soil with full treatments and sweet clover for green manure, yields have averaged 98 bushels of corn and 37 bushels of oats.

Not only have yields per acre been increased, but shelling percentage of the grain has likewise improved. Corn grown on soil with full treatment shelled 79%, compared to 70% for the corn on the untreated soil.

Full soil treatment on the fertilized plots at the beginning of this experiment included five tons of lime, 1,000 pounds rock phosphate, and 100 pounds muriate of potash. Furthermore, in years when the four year rotation plot has been in corn, some 300 pounds of 3-12-12 starter and 100 pounds of nitrogen has been used per acre. Additional potash and phosphate were required after the first round of the rotation—with the four-year rotation requiring more of these two elements than the two year rotations.

Of the small amount of rainfall received in 1953, nearly one inch was lost in runoff from the plot in corn on untreated soil. With full treatments the loss was only one-quarter this amount.

Average rainfall during the corn season has been nearly 20 inches during the past six years. Of this amount, less than three inches appeared as runoff—causing 2.25 tons per acre soil loss—under corn on the untreated soil. With the two year system receiving full soil treatment, runoff was slightly less than one inch and erosion less than one ton per acre. With the four year system including full soil treatment, runoff was a little more than one-half inch and erosion was close to one-half ton per acre.

WHAT TEXTURE OF SOIL IS PREFERRED?
I: SANDY SOILS WORK EASILY

To the question of which soil is preferred we usually get the reply, "Sandier garden soils are preferable. They are easily tilled without tearing up the tender seedlings. They are warm in the early spring. They are not hard and cloddy when dried out, and they do not pack down on tramping or under power tools going over them."

This illustrates our inclination to think first about how easily the tillage job can be done. Seldom does our thinking raise the question as to how well the soils produce in quality for our nutrition and health as well as for yields. Sandiness of the soil should quickly provoke that question.

Soil texture needs to be viewed in its variation; this includes (a) the sandy soils, (b) the silty soils, and (c) the clay soils. This variation in physical properties causes diversity in the soil's potential productivity. In our preference for sandy soils, because of the ease with which they can be tilled, we fail to see the common deficiencies in readily available fertility, to say nothing of deficiencies in fertility reserves. We cannot buy our better vegetables of high nutritional values without significant labor as to tillage, and other labor phases of soil management.

When we speak of soil texture, we refer to the percentages of the coarse, the medium and the fine mineral particles composing the soil. We recognize and estimate amounts of these by the "feel" of the moist soil when it is rubbed between thumb and forefinger. The coarse, or the sand, particles are felt as individual grains. The medium, or the silt, particles feel smooth and velvety. They are too fine to register as single grains in contact with the skin, but between the teeth they are recognized with pronounced exaggeration of their size. The clay gives a non-grain but sticky feel, and dries into clots. This is much less the case for silty soils and seldom so for sandy ones.

Since most cultivated soils are commonly a mixture of these three, they are more often called loams. A loam is one which is not pronounced sandy, nor

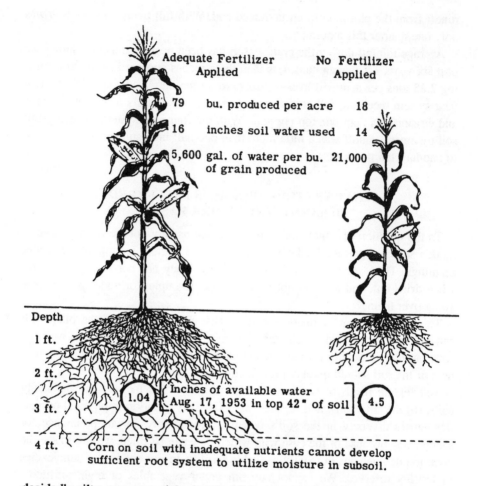

Adequate Fertilizer Applied		No Fertilizer Applied
79	bu. produced per acre	18
16	inches soil water used	14
5,600	gal. of water per bu. of grain produced	21,000

Depth
1 ft.
2 ft.
3 ft.
4 ft.

1.04 Inches of available water Aug. 17, 1953 in top 42" of soil 4.5

Corn on soil with inadequate nutrients cannot develop
sufficient root system to utilize moisture in subsoil.

decidedly silty, nor very sticky in "feel." It requires more than half of the soil's
body of sand to make it feel sandy; half of it as silt to make it silty; but from
only one-fifth to less than one-third of clay to make it a "clayey" soil.

If the soil is to hold available nutrient elements like calcium, magnesium,
potassium and others against loss from rain water percolating down through,
then the sandy soils do not represent enough surface per unit weight or volume
to contain a significant supply of exchangeable nutrients for more nutritious
plants. Their holding capacity can be increased by only the extra organic matter
incorporated as a means of giving them additional adsorption and exchange
capacities.

Sandy soils are ready for tillage and plantings earlier in the spring because
they do not stay waterlogged very long. Consequently, they warm quickly, do
not "puddle" when worked, and their organic matter undergoes active decay

early and rapidly to give both inorganic and organic nutrition to the crop. Because sandy soils need organic matter, gardeners on such soils take readily to using compost because the effects of such reflect themselves quickly and decidedly in the improved growth of the crop.

Sandy soils of low organic matter, and thereby low in absorbing-exchange capacity, make one respect the possible damage to seeded crops from salt (commercial) fertilizers remaining longer as salt solutions instead of by being made less active as salts be being adsorbed on the extensive surfaces of the ionic and finer colloidal minerals like the clay.

Plants do not use all the salt fertilizer at once, hence it must be held over for some period of time. In sandy soils this is but a short period, and much of the purchased plant foods is lost in the percolating rain water unless the applications are repeatedly made as small dosages during the growing season.

Sandy soils in the humid areas of higher rainfall do not represent much exchangeable fertility on the surfaces of the particles. Nor do they represent much as reserves in the undecomposed minerals. Particles as large as sand-size are made up of the mineral quartz. This is about the hardest and most insoluble of all the minerals. For this reason it remains as the larger leftover sizes from a natural original mixture of many minerals after all the others have weathered away. For this reason sandy soils in regions of ample rainfall, generally, are soils with much "site" value but less sustaining or productive power, unless they are built up in organic matter for the favorable action of which the sand is mainly the retaining framework or skeleton. Quartz, made up of silicon and oxygen—the former a non-nutrient so far as we know and the latter not used from that source—offers nothing as plant nutrition.

However, sandy soils in the arid areas of lower rainfall may be pulverized, well-mixed, unweathered rocks of widely varying mineral contents and much potential productivity when given water to decompose them and to make their products of that breakdown ionically active. Sandy soils have a favorable physical condition for easy tillage, but whether they are productive depends—like with any soil—on the mineral fertility reserves they have, giving—by decomposition—the necessary adsorbed—exchangeable elements, held on the finer particles of either or both the clay and the organic matter for service in nourishing the roots of plants. Soils, like any body of life, need a good anatomy, which sand can give; but with that there must go the complete physiology by which that body functions or performs in growing crops worthy of their classification as food or feed of high value.

SILT LOAMS—NUTRITIONAL BLESSING OF THE WINDS

Any single positive answer to the question "What texture of soil is preferred?" may well be adroitly avoided until we appreciate what else may be hidden by, and associated with, that physical property. It is, nevertheless significant to note that on the silt loams or soils of medium texture in the mid-continent of the United States, there is the highest percentage of the land in farms.

The wheat belt of the United States is covered with silt soils. Prairies and plains in the southern hemisphere are also of medium texture, or silt loams. It was over soils of this physical makeup that the American bison roamed. It is on soils of this texture that we grow most of our beef cattle, sheep, pigs and, not so long ago, many fine horses. It is on the silt loams that the grasses established their reputation, not only as cover against erosion by water and wind, but also as the nutritious feed for growing animal muscle, for protecting animals from diseases, and for encouraging fecund reproduction.

Soil texture determines the soil's suitability for such matters as ease of tillage, intake of water, aeration, drainage and others connected readily with physical forces. But there is much in soil texture reflecting the experience of the minerals on developing into soils under the climatic forces and consequent chemical activities. These forces include total rainfall, temperature, winds and other meteorological vagaries by which the silt loams, for example, became the surface soils over extensive areas in certain climatic settings.

It is not so much by texture, as by highly-mixed, less-weathered and much diversified mineral contents transported from less humid to more humid areas, that the silt loams of our mid-continent have been of high nutritional value in our fields and gardens for our animals and ourselves.

Silt loams are usually deposits by the wind. Consequently, to be picked up by the air currents the mineral particles must be small; they are also separated from other particles, and must therefore come from a region of low rainfall. Little clay has been developed at their point of wind pick-up, otherwise the mineral particles would be collected into granules too large to be wind-borne.

On drier areas with hot days and cool nights the wind currents are apt to be stronger and pick up particles too large to be carried when the currents slow down. By picking up and dropping fine mineral particles, the wind is a winnowing process which is carrying and segregating the minerals from our drier West to the less arid or more humid mid-continent. Silts are, therefore, on an eastward march in dust storms and are not viewed as a blessing because they rob the West of fertility rich silt to scatter it over the semi humid mid-continent.

The flood plains of the Missouri River, especially along its eastern quarter or more, represent fresh western materials from which fertility laden silt can be winnowed out, carried away by the southwesterly winds, and deposited over

Iowa, northern Missouri, Illinois and adjoining states. Deep deposits of loess along the bluffs at the Missouri River testify to such origin. Food plains of the Mississippi River serve similarly.

Experimental dust collections and study in the line of the prevailing winds to the northeast of the Missouri River for more than 15 years have shown the annual deposits to range from 750 to 1,000 pounds per acre. One single dust storm in April 1936 deposited the larger amount in 24 hours.

The mineral contents of these carefully examined deposits were commonly about 40% other-than-quartz to suggest that they had undergone little decomposition. They were a natural deposit of reserve mineral fertility to contribute to nature's way of fertilizing the soil. Nature applies pulverized mineral fertilizers but not soluble salt treatments.

Such a natural procedure deposits the reserve fertility on the soil's surface, where the weathering forces of solution, oxidation and carbonation are most active in readily mobilizing the essential nutrient elements out of the rock fragments of silt-size. Those weathering forces are more active than one commonly appreciates. They are what restores the productivity while there are no crops growing or while "the soil is resting," as we say. If a soil has ample clay from which the exchangeable or active nutrients have been exhausted by cultivation through the time of the plant's exchange of hydrogen or acidity to the clay, then the added silt minerals can be broken down by the acidity of the clay which serves as the agency for their decomposition. Resting a soil will restore productivity only when the silt is made up of undecomposed minerals, or those just recently added and undecomposed.

If, therefore, silts coming from the original magmatic granite of the arid West are nature's windblown fertilizer on the more humid soils farther east,we may well raise the question as to whether or not we might wisely apply granite dust on our soils as reserve fertility or minerals to be mobilized by the acidity of our soils? This would seem a very appropriate question for garden soils or others intensively cultivated and maintained at high levels of decaying organic matter.

Not because of texture only, but because of their contents of rapidly weathering mixtures of minerals carrying all nutrient essentials, the silt loam soils are much preferred where the crops produced are to be of high nutritional value. Our gardens dare scarcely be other than that. They are a blessing if they are made up not only of the texture so favorable in the soil's physical properties but also in its fertility or creative properties for living things as well.

WHAT TEXTURE OF SOIL IS PREFERRED?
II. CLAY—THE SOIL'S JOBBER

Clay is apt to be considered the "dirt" of the soil. Clay is sticky when wet. It is also plastic and easily molded. Then, too, it can be slippery, as one knows whose car has gone off the pavement and into the muddy shoulder where the wheels find little traction.

Such an indictment of soil, as "dirt" or "mud," emphasizes the properties by which it soils the hands, or clothes, or the car. But while these features may seem derogatory in casual thinking, they are the very properties by which the soil holds water and fertility elements as drink and nourishment, for the growth of plants and thereby for our nutrition.

Since clay is a mineral of such extremely small particle size, to make it behave according to the surface of the mass rather than its weight (2.65 times as heavy as the water it displaces), it is called a "colloid." That word means "glue-like." Clay is sticky when wet, but hard—almost like cement—when dried, hence the propriety of the term.

The behavior of holding water strongly enough for a slowly drying process gives clay a big value in holding a very thin film of water between the plant root and the clay particles of the soil around it. It is within that very limited thickness of a water film around the clay particle that the clay gives its important services as the adsorber or "taker-on" of nutrient elements, should they come along in solution from decaying substances, rock particle decomposition, or from applied fertilizers. It is, then, from that thin film, holding nutrients against loss to the leaching rainfall, that the clay is the exchanger or "giver-up" of nutrients. They are taken from there, however, by the plant root only as it exchanges for them correspondingly "active" elements in the form of their ions, most commonly the non-nutrient hydrogen or acidity. Because of the clay's immense surface in its finely divided, plate-like particles, it is an adsorber and exchanger—a jobber, not a manufacturer—of ionically active elements coming out of rock minerals as recombinable parts so separated and activated in the presence of water.

While the two soil separates, namely, sand and silt, do not soil our hands and are not commonly considered "mud" or "dirt," they do not have the significant capacity to filter out, or to absorb, nutrient elements from solutions of them. Solutions applied on sands are apt to leach away to the sea. Sand and silt soils are more quickly responsive to fertilizer treatments, as shown by the plant growth. However, the fertilizer effects don't carry over into the following years.

Clay soils do not give response to fertilizer applications so pronouncedly by marked crop manifestations. But on these the salt danger from fertilizers is not so serious. Nor is salt damage so severe. Some of the active fertilizers may become "fixed." That means that some element may be held by the clay so firmly that the crop manifests no growth response therefrom. It is these speedy and pronounced performances that mark out the clay for (a) holding the solubles on itself in adsorbed-exchangeable form against loss of percolating water; and (b) for giving them up when some active soluble from the root or otherwise is exchanged for them. In these respects, the clay is the textural separate of the soil serving as the jobber of essential plant nutrients in their active forms.

When the nutrient elements are first broken out of the rock while it is being weathered into soil, the clay, as a secondary mineral, is forming from the less soluble rock remnants, and immediately enshrouds even the smallest decomposing rock particle as a coating. This clay cover modifies the behavior of the remaining rock components in their decomposition and development into soil.

The differing rates of the weathering of the rock-minerals, or of the development of soils from them, under different climatic forces, means different properties of the resulting clay. Under higher temperatures and higher rainfalls, it may have proportionately less silica. It will have correspondingly more oxides of iron and aluminum in its own chemical-mineral structure.This gives it less adsorbing-exchanging capacity. It will hold less "available" nutrients for crops. It won't hold as much acidity either, consequently it does less business as a "jobber."

It was the refined study of soil acidity under research some 20 and more years ago that revealed these basic facts about the clay as the result from rock weathering, and its role in plant nutrition. Now that we have learned that a certain clay developed in a certain climatic setting has its own particular capacity of nutrient adsorption and exchange per unit weight, we can measure the soil's fertility capacity for different ions. Since these are represented mainly by the positively charged nutrients, namely, calcium, magnesium, and potassium (filling almost 90% of the clay's capacity together in the approximate ratios of 75, 10 and 5%, respectively) in a productive soil, then when once the soil becomes highly loaded (over 10%) with the non-nutrient hydrogen so as to be called an "acid" soil, we know that the trouble may not be due so much to the presence of the hydrogen in the jobber's adsorbed stock of ions as to the shortage there in what would be these other positively charged ions serving as active and available plant nutrients coming from that source.

In getting a soil ready for the planting, its higher content of clay may delay the crop season while waiting for the soil to dry. The clay may make it workup in big lumps, or give hard dry clods calling for more work before getting into seedbed condition. There may be other labor aggravations because of the clay in the soil. But when one comprehends what clay will do in a nutritional way for plants as a jobber of active essentials, if we will do our part, we shall certainly find the clay doing its part.

WATER—NATURE'S MAJOR BIOCHEMICAL REAGENT

We commonly take for granted the water we need since so many of us live in the humid soil regions. Those are areas where the rainfall (inches per year) exceeds the evaporation from a free water surface (measured similarly). We may use water for irrigating plants in the garden or field, but for most of the agricultural crops of our country we depend on natural water in rainfall. We think of the mass of water as so many gallons or acre-inches, and of its physical properties related to temperature. We do not recognize its extensive and vital significance as a biochemical reagent in the growing garden and field plants and in all living warm blooded bodies.

Water is the first major mineral which our plants take from the soil. It is combined chemically with the atmosphere's supply of carbon dioxide. The process gives the major synthetic product of plant growth, the sugars and the initial process for conversion into other products. Sugars, and several organic compounds chemically similar to them, are called carbohydrates. This means carbon plus hydrates or water. These two are combined in equal ratios. The carbohydrate compounds lose water when heated. Heated sugar first becomes caramelized. With more heating it becomes black carbon. On final burning in the air this abundant organic element gives off a large amount of heat. This makes the sugars the major energy giving compounds in the plant's biochemical processes, just as sugars, starches and carbohydrates are the major energy foods for our animals and ourselves. They represent the sun's chemically stored energy to be digested and released within the plant tissues (similarly in our bodies) when oxidized or burned in the plant's biosynthetic processes, or those carried out by life for its own maintenance.

Water combines with carbon dioxide at the ratio of six molecules of each. This gives a simple sugar when the leaf carries on its photosynthesis under the energy supplied by the sun and the catalytic help of chlorophyll.

Our emphasis on the amount of water required to grow crops is a natural emphasis, when plants pile up their carbohydrates so rapidly as increase of bulk. In using irrigation, we choose to grow those crops composed mainly of sugars, much vegetative bulk and watery or juicy contents. For it is only by marketing such quickly produced and highly carbohydrate products that the requisite larger monetary returns, by sale of the costly irrigation water, can be realized.

Other elements of mineral origin are usually not so plentiful, nor so active and well-balanced in the partially developed, often sandy soils. In such soil regions, the water is the major biochemical reagent in food synthesis, and the food products are mainly sunshine, air and water, or energy foods. The biosynthetic performances by plants of converting the carbohydrates into proteins are not so prominent under naturally arid conditions, or on those soils given only water as soil treatment by irrigation.

Fats in the seed or plant, as stored energy, are also made by chemical reduction of sugars. Fats are digested in the plant's use of them by processes (as in our bodies) in which the water, a mineral, is again the major biochemical reagent. We have not emphasized water enough in its biochemical importance to realize that this liquid mineral is the major reagent in organic and living processes, much as it is the major mineral among the other minerals contributing to nutrition (via the soil) of plants that synthesize energy foods for their own growth and the growth of all else that is nourished by them.

When we see water doing so much to give large crop yields of energy foods we need to remind ourselves that even energy foods from photosynthesis cannot be produced except by the living substances, namely proteins, and we must remember that proteins are not synthesized directly by the sun's energy, but by life processes requiring sugars for energy and for starting compounds, not forgetting all the soil fertility originating in minerals other than water. While photosynthesis may be using sunlight with an efficiency of 30% in piling up carbohydrates as bulk and energy values, the biosynthesis of proteins as living substances represents the use of sunlight with an efficiency of only 3%. The plant, much like an older, fattening animal, gains weight rapidly by piling up stored energy in the carbohydrates, but its protein production is slow, like a young animal growing muscle and living tissue.

Plants combine the simpler elements into compounds by using the sunlight. Plants alone are unique in this respect. By it they make energy food and protein food values on which man and animals are dependent. Microbes are also dependent on the plants for carbon or energy food values. By decomposition of organic matter for energy, the soil microbes again use water as a soil mineral to decompose plant remains. Thus they set free carbon and all the other elements to let them make another cycle of service in growth and decay again plays the role of hydrolysis, etc., in the biochemistry of microbes as it does in other life forms. Carbon released as carbon dioxide by microbes is then united with water to make carbonic acid, nature's most universal acid for decomposing rocks and their constituent minerals for plant nutrition.

In the soil, water is the mineral giving major chemical and biochemical services, seemingly unheralded and unsung. Perhaps a better understanding of water will do much to help us manage soil for the production of crops to better advantage of these in our own nutrition.

A WEAK SOIL BODY

When soils erode, we are apt to take up the fight against running water. This is much like when some disease comes over our body, we think first about fighting the microbes. When we break a bone we put the limb in splints. Similarly, when a field is broken up by gulleys, we line it up with terraces.

Whether it is the soil or the body that is in trouble, we fail to realize the deep seated, preceding but gradual, weakening of our body or bones and of the soil body too. That occurs long before the noticeable disaster of the fracture or gulley befalls us. Broken bones too often are the result of malnutrition ahead of them to make the bones weak. Coffee and toast don't maintain strength. Unsteadiness in muscle may have come along with the weakening skeleton to bring on the fall, as well as the weak and broken bones. In like manner, the exhaustion of the strength of the soil, its fertility, weakens the soil body and makes erosion the consequence.

That such are the facts for the soil body is suggested by the experimental plots on Sanborn Field at the Missouri College of Agriculture. That field, after 62 years of its recorded behaviors, is a sage to tell us what the experiences of the soil body mean in what can be "old age" of it. Two plots have been planted to corn each year since 1888. Professor J.W. Sanborn outlined the use of six tons of barnyard manure annually on one of these, while the other was expected to go forward in crop production with no soil treatment. Fortunately these two plots are alongside each other. There is a good sod border on three sides, or in the direction water might run on these seemingly level areas. All of the crop, grain and fodder, is removed outside of the return of fertility in six tons of manure on the one plot. The management and history of these two classic soils have been exactly the same.

That the removal of the fertility without return on the no treatment plot has weakened the soil body to make it erosive is now clearly evident. Had the sod border not protected this plot, its soils—like so much from the rest of Missouri—would now be resting in the Gulf of Mexico near New Orleans. After that soil body is turned by the plow, a single rain is enough to hammer it flat, to seal over the soil's surface to prevent infiltration of the rainwater, and to bring on erosion of that fraction of the surface so highly dispersed by the raindrops. This was recorded by a photograph of this plot taken one Friday in spring before plowing for corn, and followed by another photograph on Monday of this plot taken one Friday in spring before plowing for corn, and followed by another photograph on Monday after the plot had been plowed on Saturday and a rain came on Sunday. The plow-turned form can be completely obliterated by one rain in a weak soil body.

Where manure had been going back regularly each year, naturally there was a different soil body. It stood up under the rain and maintained its "plow-turned" condition in spite of the rain.

It was the same rain that was so "damaging" to the other plot. One could not blame the rain for any damage here on this manured plot. Instead, the rain brought benefit. Its water went into the soil. It soaked a deeper soil layer and built up the stored water supply for the summer. This soil is cooler by ten degrees in the summer than the companion plot. Here is a different soil body that behaves differently under the rainfall. It doesn't erode. The rills of running water began at the line that divides the two plots. Narrow as these plots are, there are rills on the "no treatment" but not on the "treated" one. The former might seem to be a call to "fight" the running water. The latter is not.

Fortunately the "strength" of the soil body against erosion in this case is also the "strength" of the soil for the crop production. It is also the "strength" for soil granulation. The corn yield is still twice as large on the plot with manure as without it. Weeds grow on the former after the corn roots are deep enough to be beyond the use of the nitrates accumulating in the surface to invite the seeds. These weeds are a nice winter cover and they are one that comes there without cost to the management. The granulation of the soil of the manured plot is so much better under laboratory test than that of the unmanured one, that water goes into the soil three times as rapidly. Also it moves about four times as much volume of water down through and does not plug itself up quickly to stop water movement in the soil.

Here is "strength" of granulation. It is the "strength" of the soil body under the hammering effects of the falling rain. It is the "hidden" strength, and the very same strength, namely the fertility that gives the bigger yields of crops. This is distributed within the inorganic as well as the organic fraction of the soil.

Here is quiet testimony that we ought to see the weak soil body and the erosion of it brought on, because we have removed the fertility, or the creative power, by which any soil naturally keeps itself in place and grows nutritious crops at the same time.

Our weakening soil body is suggesting gradually weakening human bodies created from it.

3

SALTS FOR THE
WOUNDS OF AGRICULTURE

TOO MUCH SALT FOR THE SOIL

Plant growth is disturbed by small concentrations of salts in a soil not only in alkaline or low rainfall areas. Similar disturbances are becoming more common in the humid, or higher rainfall, regions. We are putting the western deserts on the march eastward, as it were, by heavy applications of highly concentrated fertilizer salts.

Garden soils with their larger amounts of organic matter, and the increased microbial activity which that fosters, are not so susceptible to visible "salt-shocks," like the defective germination of seeds and the poor stands of the crops, as has been the increasing experience in heavily fertilized fields not regularly manured.

When nature demonstrates more abundant plant growth under rainfall high enough to leach salt solutions out of the soil, the remaining nutrients are held there by adsorption on the clay and humus colloids. Those essentials go to the plant mainly when the roots exchange hydrogen, or acidity, for them. From that illustration of natural crop growth, we may well respect the common warning of a farmer or a gardener when he says, "You can easily use too much of those chemical salts."

That warning may come from some unfortunate experiences, possibly from putting the fertilizer salts in the row with the seeds and getting no stand. It may come from the disturbed flavors and tastes of the vegetables, like radishes, spinach, onions, and cabbage. It may come even from more plant diseases and more insect troubles.

The pioneer farmer was cautious about using salts, as illustrated by his reluctance to accept Chile saltpeter after he had used bird dung or guano from Chile. Even Dr. J.W. Sanborn, who laid out one of the oldest experiment fields of the country in 1888 in Missouri, had nearly half of the plots include barnyard manure as the soil treatment. He ventured only one plot of continuous wheat with significant applications of commercial fertilizer salts alone. He used those to put back the equivalent of the nitrogen, phosphorus and potassium taken from the soil in the grain and the straw of a 40 bushel crop.

The history of the experiences with that plot under chemical salts now for 68 years tells us the salt injury story. The harmful dosing was not recognized as serious or as a disaster until after about 20 years of continual dosing. Then the drilling of that much salt with the seeding resulted in poor emergence of the seedlings and poor wheat stands. If one interprets the records carefully, it required a few years of seeding with such dire experiences to suspect the salts and not the seeds as the cause of crop failure. Subsequently the records report that the application of the fertilizer was reduced to but one-half the total amount drilled with the October seeding. The other half was applied as a top dressing in the spring. Later in the plot's history, when fertilizers became more concentrated, even that reduced amount became more dangerous to the stand of wheat and still lesser amounts of salts were drilled with the seeding.

The average yield of wheat on that plot fertilized heavily for now over 65 years has not been near the 40 bushels. It has been only slightly over half of that. Only nitrogen, phosphorus and potash, along with any contaminant or extra elements, were not maintaining the yields. Later when the more common trace elements and salts of some neglected others were added, the yields showed their improvement even though the organic matter in that plot is one of the lowest among the many on the field.

When the growth of plants in solutions of salts in the demonstrations by a botanical class, or in commercial hydroponics, calls for such very dilute salt solutions—and those renewed often and regularly—we ought to appreciate and respect, (a) the dangers from salts in the soil, and (b) the "shock absorber" services by the microbes along with the organic matter and the clay to take the ionic salts out of activity before they injure the seedlings of the plants. Also we might recognize all the commercial activity in designing new farm machinery for particular placement of fertilizers away from the seeds in the soil. This is testimony that (a) with more concentrated salts applied, and (b) with the organic matter content of the soil decreasing to reduce the microbial activity, there isn't enough "buffer capacity," or "shock-absorbing" ability left in the soil to remove the hazards from the "salt" effects.

Our soils are becoming sick and weak. Unlike the patient encouraged by the doctor who says, "You have a good constitution," we cannot say that for our soils, if we define that property as the ability of the soil to produce crops and maintain its naturally healthy condition in spite of the soil doctors rather than because of them.

The wise gardener or farmer will give his soil a "strong constitution" by using not only vegetable wastes—either composted or not—but also animal manures for the sake of their organic matter of direct plant origin and also for possibly some of the distinctly animal origin. If soils required additions as fertilizers to correct some of their nutrient deficiencies, the "salt effects" may be buffered by their use through a composting process either in the pile or plowed under as application with much organic matter. We need to realize that we can use too much salt on the soil.

BUYING FERTILIZERS WISELY

Commercial fertilizers are carriers of nutrients for plants. They are designed to be applied to the soil for improvement in crop growth. They do not contain all the chemical elements required by plants, but only those commonly deficient in most soils for better yields of crops. In the purchase of fertilizers as means of getting better crop yields, it is essential to be familiar with, (a), the kinds and amounts of plant nutrients the fertilizers usually contain, and, (b), the effort of the inspection service of the state and of the fertilizer producers in supplying farmers with efficient fertilizers.

Most commonly, fertilizers carry one or more of three chemical elements essential for plants, namely, nitrogen, phosphorus, and potassium. These are listed on the label of the fertilizer container as nitrogen (N), phosphoric acid (P_2O_5) and potash, (K_2O). Other essential elements may also be carried by the fertilizer. They may be included in the description of it on the label. Calcium, and magnesium, together with some other elements commonly spoken of as "minor" elements have been more recently given attention for their values in fertilizers.

Because the nutrient elements, common in fertilizers, are combined with other elements into chemical compounds, and do not occur singly for such use, it is impossible to have a fertilizer made up of nitrogen, phosphorus and potassium which total 100%. These chemical compounds, called the "carriers," are put together to make the fertilizer mixture. Considering all the three nutrient elements in combination (nitrogen, phosphorus, and potassium), a fertilizer is fairly concentrated when, in total, it carries as much as 20% of those combined nutrients in forms which are stable and serviceable to the crop.

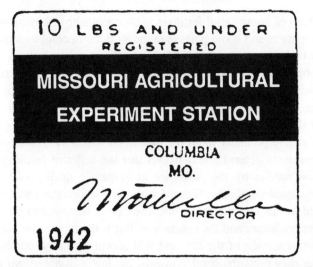

10 LBS AND UNDER
REGISTERED

**MISSOURI AGRICULTURAL
EXPERIMENT STATION**

COLUMBIA
MO.

DIRECTOR

1942

That fertilizers should contain mainly three chemical elements when at least 14 are required to grow plants is due to the fact that these three are most commonly deficient in soils. Their application is most commonly effective in giving better plant growth. Nitrogen is needed to form the protein in the plant or the basis of cell multiplication as growth and life itself. It is especially needed in the early life of the plant, it is the most costly element and the quantity in the fertilizer is not relatively large.

Phosphorus like nitrogen is needed for growth. There is usually enough in the seed to give the plant its start but the roots of the young plant must soon be taking the phosphorus from the soil. Phosphorus is especially important in producing the seed and therefore plays an important part in the early maturity of the plant and in the yield of grain. Phosphorus is lost from the soil when grain and the bones of animals are sold from the farm, and must be returned to the soil largely by the addition of fertilizer. Phosphorus makes up the bulk of most mixed fertilizers.

Potassium serves particularly in carbohydrate production. Taken from the soil by the plant it helps the plant to utilize the air, water and sunshine which produce the body or mass of the plant. It is removed from the soil by the hauling away of the stover and straw of the crop. Unless the soil is one in which this element occurs naturally in large quantities, it is usually only slowly available to the plant. It must be present in ample supply and most generally be supplied as water soluble potash salts in the fertilizer applied.

To sum up, nitrogen, phosphorus, and potassium are the three elements that are used in large amounts by the plants and are limited in the soil supply as available forms during the growing seasons. Therefore they are commonly applied and represent the plant nutrients of main concern in the fertilizer business.

The purchase of commercial fertilizer is thus a matter of buying some nitrogen, some phosphorus and some potassium, or all three in combination, as soil treatments for better crop growth.

The cooperative efforts of the fertilizer producers and the inspection service of Missouri in supplying farmers with efficient fertilizers dates back to the enactment of the Missouri Fertilizer Law. This came about through the activities of the groups in the state interested in the promotion of agriculture. The law aims to give publicity to all the factors on which the values of the fertilizer as plant nutrients depend and demands that the seller of fertilizers give all the information needed by the purchaser to judge the quality of the goods. The purchaser should familiarize himself with such information and understand the terms used in labeling the fertilizers. The law is designed to be fair protection to both the producers and the consumers. But if the buyers are not familiar with the protective service of the law and will accept goods on purchase regardless of whether they meet the legal demands, the law will not afford the protection intended. In its service to purchasers, the law requires (a) that the fertilizers be registered with the state as to plant nutrient contents and name; (b) that the containers of the fertilizers bear tags certifying to this registry; and (c) that the containers bear labels giving the plant nutrient contents. The fulfillment of these phases of the law is a guarantee, by the inspection service, of fertilizers sold in the state. It is essential, therefore, that the purchaser be familiar with the various phases of fertilizer control for him, through the Missouri Fertilizer Law.

The requirements of the law include the following:

1. Registration: Every manufacturer, importer, or other person or company responsible for placing any fertilizer, or material to be sold as a fertilizer, on sale in the state must file the name and the address of the manufacturer, the name under which the fertilizer is sold and its guaranteed chemical analysis.

2. Tags: The fact of registration is made known to the purchaser by the presence of a registration tag attached to the bag. The form and information on this tag as used in 1942 on bags weighing more than 50 pounds and up to 100 pounds is given on the following page.

Such tags shall be placed in plain sight of the purchaser on all the bags within this weight limit which are filled and leave the factory during the calendar year 1942. New registration must be filed and new dated tags obtained for each year.

The buyer therefore should look for this tag or label on each container purchased. If it cannot be found, the package or sack should not be accepted.

The date of the tag or gummed label on the package may be of some previous year. This does not mean necessarily that the fertilizer has lost any of its value in storage through deterioration on standing. Fertilizer which has been

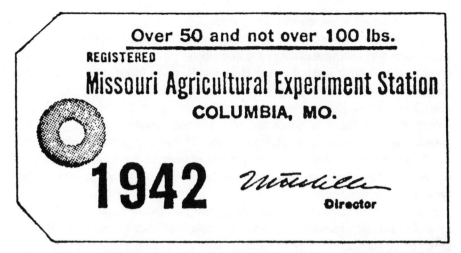

Over 50 and not over 100 lbs.

REGISTERED

Missouri Agricultural Experiment Station

COLUMBIA, MO.

1942

Director

kept dry in the warehouse, or has not been exposed to excessive heat, is as good or better than when it left the factory. Much of the tendency of freshly mixed goods to pack or stick together disappears during storage. Storage may improve the drilling quality of it. The presence of a registration tag of former date, therefore, does not constitute a reason for condemning the fertilizer.

Every handler and seller of fertilizer should keep in mind that the registration tag accompanies the fertilizer, not the bag. If broken containers necessitate placing the fertilizer into new bags before being sold, the registration tag on the old bag should be placed on the new container into which the contents have been transferred.

Sometimes goods leave the factory without registration tags attached, and the registration tags are sent separately by mail or otherwise to the receiver. If the receiver is to sell the goods, these registration tags must all be attached to the several bags. It is not sufficient that the required number of tags be handed separately to the buyer. Bags with the tags not attached are not ready for legal sale and should not be received by the buyer.

3. Labels: Every bag, package, box or container containing fertilizer offered for sale must be labeled as to the fertilizer contents. This label may be stenciled or printed on the outside of the container or it may be printed upon a tag attached to the same. If printed on a tag, it must not be printed upon the registration tag (described in the previous paragraph). The label and the registration tag are to be distinct and separate.

The label on the container must state the name and address of the manufacturer, the name of the fertilizer brand, and the guaranteed chemical analysis of the fertilizer. The guaranteed analysis as it appears upon the label must state (a) the percentage of nitrogen; (b) the percentage of available phosphoric acid; and

(c) the percentage of potash soluble in water.

There are some exceptions to the above requirements as to the statements of guaranteed analysis. In the case of ground bone meal, either raw or steamed, dried meat and bone tankage, dried meat scraps, dried blood, and blood meal, the percentage of total phosphoric acid is stated instead of available phosphoric acid. These materials, which carry their phosphorus as that of animal origin furnish the only cases in which the words " total phosphoric acid" may be substituted for "available phosphoric acid." If the words "total phosphoric acid" appear on the label of any other fertilizer material the words "equivalent to" must precede them, and the statement "available phosphoric acid" must be present also on the label.

In all cases where phosphoric acid is present in a fertilizer and is guaranteed as to the amount present, the words "available" or "total" must precede the term "phosphoric acid" in every case. In other words such a statement as:

> "Phosphoric acid ____%"

cannot be accepted as a satisfactory label.

For materials which do not contain all the three fertilizer constituents, nitrogen, available phosphoric acid, and potash, only those constituents which are present will be mentioned in the guaranteed analysis.

Fertilizers, like other commodities, are purchased most wisely when their service is high in relation to their cost. In order to use fertilizers most effectively they must meet the shortages in the soil. Their use in a small way as a first experience is to be recommended until one knows more about their service in connection with the soil and the scheme of its management in question. Goods of high quality usually render fullest service.

Fertilizers not only increase the yield, but they may render service in the improved feeding values or quality of the produce. Because of high values of farm products, the needs for better nutrition in animals and humans, and the relatively low costs of fertilizers now as aids toward maintaining the soil fertility, the purchase of commercial fertilizer deserves careful attention and wider adoption as farm practice.

1. Which company sells the best goods in Missouri?

ANSWER. Any company which complies with the conditions of registration, labeling, and putting on the registration tags is a dependable firm with which to deal. By consulting the annual report on fertilizer inspection, the stand of each company for each and all kinds of fertilizer sold, expressed in percent on the basis of the analysis of their goods, can be found. The analysis of each company's separate samples is also reported there. These data can also be used in judging the cooperation of producers to maintain fertilizers as labeled or guaranteed.

2. What kind of mixed fertilizer or superphosphate shall I use for a particular crop?

ANSWER. The Missouri Agricultural Experiment Station will suggest a list of fertilizer mixtures or superphosphate to be used for different crops in various parts of the state. Their recommendations are based upon actual experience and experimental tests, conducted on many of the different soil types of the state. The list of fertilizers recommended for general farming in Missouri includes, 0-20-0 (superphosphate), 0-20-10, 1-20-20, 4-16-4, 4-10-6 and 10-6-4. Nitrogen carriers such as ammonium sulfate, sodium nitrate, or calcium cyanamid may be used in conjunction with the fertilizer if more nitrogen is needed. Other analysis approaching the above may be substituted, such as a 4-12-4, or a 4-10-6 for the 4-16-4, or an 0-12-12 for the 0-20-20.

It is always safer to follow these suggestions than merely to use the cheapest brand of fertilizer one can buy.

3. Some companies claim that their fertilizer contains portions of the rarer elements which are needed by plants and therefore will give additional yields because of their use. Is this true?

ANSWER. It may prove to be true, if the soil in question is lacking in these elements and if the crop suffers because of this fact. On the other hand if the soil is supplied with these elements in sufficient quantity for crop needs, the addition of more of these will not increase the fertilizing effect; in fact there might arise a condition of injury to the plants because of more being added in the fertilizer. The value of these cannot be predicted in any particular case. Only by trial in the same year with the same crop on the same field using the rare element fertilizer, ordinary fertilizer without these rare elements, and no fertilizer whatever, can this question be answered.

4. So much is said about lime and phosphate as though these two go together. Must lime be used with fertilizers?

ANSWER. For Missouri soils, liming for legumes has long been recognized as essential. Phosphates have been widely used on small grains which serve as nurse crops for the legumes, and the benefit from the phosphates has carried over to the legumes. Since legumes are the first means of soil improvement in fertility, naturally they have brought limestone and phosphate together. They really belong together so far as the soil-building effect of the legume crops goes, since phosphate applied on a lime-deficient soil is not recovered as effectively by legumes and even by non-legumes as when used on a soil not deficient in lime. Unless the plants' need for lime is met by the soil, much of the investment in phosphate fertilizers is not recovered. Lime and phosphate really go together in their service to all crops.

Fertilizers as a whole are more effective when used where plants are not suffering lime shortage. This does not mean that soils must be limed so heavily as to be made neutral. In many cases, drilling the limestone like a fertilizer will be sufficient. In practice it may be well to consider drilling limestone and fertilizers together. In the purchase of fertilizers, it is well to plan to use them on soils not seriously deficient in lime, if the purchase is to give largest returns.

5. Is fertilizer a good substitute for manure?

ANSWER. Fertilizer use should not serve to divert attention from manure conservation, its maximum production, and its wisest use. All possible practices in better soil management should be exercised first and then fertilizers purchased and added to make up the deficiencies in soil fertility that need to be balanced for most effective crop production. Manure use represents putting back much of what came from the soil. Fertilizer use represents putting on some fertility purchased and brought from outside the farm, to add to the soil's supply.

GETTING OUR MINERALS

When we think of vegetables we recall them by name, but with little thought as to their chemical composition. When we eat spinach, we talk about "getting our minerals."

We imagine that vegetable greens, like spinach, kale, turnip tops, Swiss chard, and others are serving to bring to our digestive tract some iron, some lime, or some copper, as if these inorganic and metallic substances delivered in those forms were rendering nutritional service. Because we measure them chemically in the plant ash, or only after the plant has been ignited, we have failed to realize that chewing nails, copper wire and lime rock, or eating them in powdered and encapsulated forms would not be the equivalent of the services rendered when these elements are brought in organic combinations by the plants. Nevertheless, even if we measure these inorganic elements only as ash forms, their amounts in the same plant species grown on different soils may vary so widely as to be scarcely believable.

In the case of simple vegetable greens, like kale for example, one might not realize that the soil treatments would give us plants looking so much alike yet so different in content of an inorganic element like calcium. In some experiments with that green, which belongs to the cabbage family of vegetables, only the clay portion of the soil was modified by having the clay filter out of solution; that is adsorb on itself different amounts of calcium and different amounts of nitrogen while it was adsorbing equal amounts of the other applied nutrients.

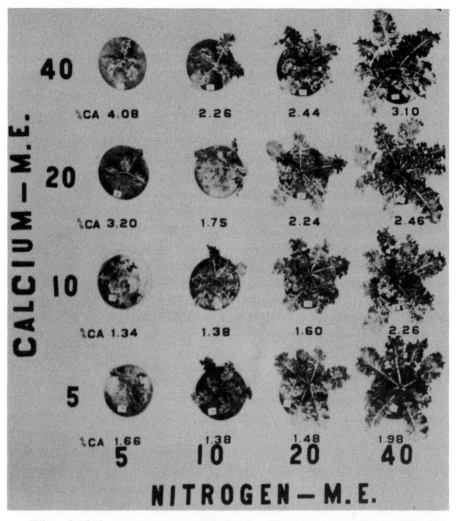

When the kale was grown to good size in some of the pots, it was still very small in those pots which were given less amounts of either calcium or nitrogen. But when the plants were put to chemical analysis for the amounts of calcium each might deliver for bone building services, let us say, it was surprising to see the wide differences in calcium content of plants that were not only good looking but alike in appearance (see vertical rows at the top of this page in accompanying illustration). The growth of kale, like that of spinach or other leafy vegetables, respond markedly to the increasing amounts of nitrogen (vertical rows, from left to right). Better appearances as well as bigger growth resulted from fertilizing with nitrogen. This better growth and good green color is the common criterion of our success as growers of vegetables and other crops.

If we could grow kale plants like any of those in the row to give 40 ME (milligram equivalents) of nitrogen, we would consider ourselves successful growers of this vegetable. That success, however, would be supported mainly by appearances and vegetative bulk produced for sale as pounds of greens. Such might not be success in terms of the nutritional value of the kale greens. That success might not stand up when faced with the questions, "How much calcium as a mineral nutrient does the kale provide us on eating it?" or "Into what organic combination is that calcium put by the plant for service in our body through ready and complete digestion?"

As for the first question, the experimental trials pointed out the variation in percentage of calcium in the dried kale, going from 1.98, to 2.26, to 2.46 and to 3.1 merely because the soils in that series were given 5, 10, 20, and 40 milligram equivalents of calcium per plant respectively. This was supplied in a form that would be readily exchangeable to the plant roots coming along to exchange their hydrogen ion, or acidity, for it.

Here, one might not be able to select one plant as different from any of the others in the group of four, yet one of these might give us 3% of calcium and the other only 2%. One plant is 50% higher in bringing to our digestive system the inorganic supplies so far as calcium goes. Looks were nicely improved by the nitrogen fertilizer. The content of the plant in calcium, however, was dependent on the amount of lime or calcium put on the soil. Looks may be deceiving in vegetables as in many other things.

As for the second question, namely the organic combination of the inorganic elements brought to us by plant growth, this is related to the plant species. It is also related to the fertility of the soil, especially the balanced or unbalanced diet for the plants offered by the soil. The spinach is a kind of plant that puts its calcium, for example, into chemical combination with oxalic acid, giving us calcium oxalate. This is highly insoluble even in acids and is thereby indigestible. Oxalates, like this one, form into needle crystals and irritate the mouth tissues as one well knows who has been treated to a small slice of the so-called "Indian turnip"; to rhubarb; or has eaten spinach that "put an edge on the teeth." Spinach grown on one combination of soil fertility for its diet may be correspondingly distasteful because of its excessive oxalate content. Some other spinach grown on other soil fertility combinations may be a pleasant taste. It is this variable fertility under the spinach, and not the plant species that has engendered love in some cases and hate for itself in others, "Popeye" notwithstanding.

Kale, however, does not put its calcium into combination with oxalic acid and does not make it so indigestible. Here a small percentage in the greens delivers much digestible calcium. This is quite the opposite of spinach. Organic compounds of the calcium into which the plant's processes synthesize or connect this essential element and not the amount in the plant ash, become the criterion of nutrient values for minerals as well as the amounts of these in the plant.

Most significant, however, is still the big fact that while plants differ as species in delivery of nutritional services through the inorganic elements they contain, it is the fertility of the soil that comes in to hinder or help in these values more than any of us are wont to receive and to appreciate. The creation of leafy plants for us still depends on the hands of dust and its creative services.

GROWTH PROCESSES EXPLAINED

The term "fertility" refers to the chemical elements contributed by the soil for plant growth. The productivity of a soil is determined mainly by its delivery of ten chemical elements in effectively balanced amounts, since these nourish the plants directly and also enable them to use four additional elements coming from the air and water. Crop production depends on the successful management of the soil so that it delivers its fertility to the crops. The fundamentals of the soil processes as they provide the raw materials to initiate and continue the manufacturing business of the growing plant, through which the sunshine power synthesizes the complex chemical compounds of vegetation, may well interest all of us.

All life depends on this natural chemical industry that draws on the soil for only about 5 or 10% of the plant's makeup, while it draws on air and water for the remaining 95 or 90%. The decline of the fertility stores of the soil is bringing these fundamentals into greater significance. It is well that we understand them while our efforts in conservation may still find sufficient of soil fertility left to be conserved.

The bulk of all plants is combustible. They are therefore made up largely of carbon, so commonly combined with hydrogen and oxygen in the ratios by which the last two are found in water. This combination as carbohydrates in their various forms is the bulk of all vegetation. When to this chemical combination there is added some nitrogen, the compound becomes protein. For legume crops, these four elements, namely carbon, hydrogen, oxygen, and nitrogen, are supplied by the air and water. They are chemically combined by the energy of the sunshine. Carbohydrates in particular represent concentrated collections of chemical energy. As sugar, starch, cellulose, wood, and others, they represent fuel values in animal diets.

The protein also represents some energy collection, but more particularly it is the growth promoting compound. Cell multiplying capacity resides in it. If true proteins are to be produced—that is, if the nitrogen, and in some cases some phosphorus and some sulfur are to be coupled up with carbon, hydrogen and

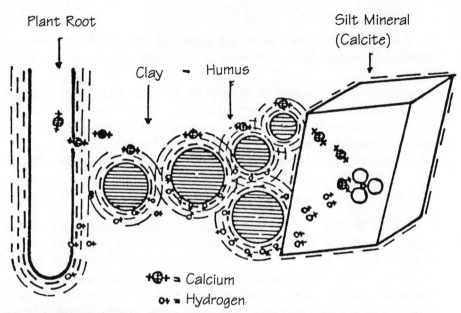

Plant Root

Clay — Humus

Silt Mineral
(Calcite)

+⊕+ = Calcium

O+ = Hydrogen

Chart 1. Diagrammatic sketch of contact exchange of nutrient ions from mineral to colloidal clay or humus and from colloid to plant root in equilibrium with movement of hydrogen in the opposite direction.

oxygen to make this life-carrying, body-building substance—they cannot be built by sunshine using only air and water. The soil must contribute at least ten elements. These ten include calcium, so common in limestone; phosphorus, common in bones; potassium, abundant in wood ashes; magnesium; sulfur; iron; boron; manganese; copper; and zinc. With the soil's responsibility toward the plant, including as many as ten elements, and with the shortage in any single element limiting the crop growth, is it any surprise that the problem of supplying the plants with their requisites may be a common one on some of our soil types?

Every plant represents a skeletal structure for absorption of sunshine, intake of carbon dioxide from the air, and absorption of water and fertility from the soil. This structure can be roughly visualized as consisting of woody material much like a building's skeletal frame. Every plant has it. Some have but little more than merely a woody framework. We can then imagine the plant's manufacturing performances in constructing this woody mass largely of carbohydrates. Water, and carbon dioxide, both represented by the air and the weather, are the sources of this material. Then, as the soil offers more of fertility to permit the manufacturing and synthesizing performances to move forward at greater intensity, the plants will contain higher concentrations of proteins, of minerals, and of other soil-borne substances to be put into abundant seed yield or larger tonnages of forage with high nutritional values for animal consumption. These are the manufacturing activities carried on within its woody frame structure. Roughly, we may view the plants as functioning to make products of value in animal body construction or in cell multiplication for their own growth from the fertility contributed by the soil. They make products mainly of fuel value from the contribution by air and water.

Vegetation can be classified, then, into two groups, the first being the woody, or the carbonaceous, group when the soil contributes little fertility and compels the plant to operate largely on water and weather. The second is the proteinaceous and mineral rich group when the supply of soil fertility is large. Forest trees grow on soils of lower soil fertility, while legumes, such as alfalfa, demand higher soil fertility. The first of these two groups reflects the fuel value, and the second the nutritional service in body building, as we all know of alfalfa's service for promotion of growth in young livestock.

Not only in the different plants are these differences found, but even within a single kind of plant there is a similar variation in composition according to the soil fertility. Soybeans, for example, become more woody in character if grown on a limited supply of soil fertility. When more generously nourished, they become rich in protein and rich in minerals as legumes are expected to be. The soil fertility supply determines the plant composition, irrespective of the plant's pedigree or its parents as performers on some other soil.

Where within the soil are the plant nutrients stored, or where is the fertility retained? This is a question that must be answered if we are to appreciate the fertility problems of some of our soils. More baffling to many is the question, how can the plant get anything from the soil after water has been going through it so long to carry soluble materials away? One needs only to recall the universal practice of burying things in the soil to get rid of odors, or of filtering water through soil or charcoal to obtain clear water, to appreciate the natural phenomenon of adsorption by the soil. Filters are made of substances offering extensive surfaces by which materials in solution are taken out of solution and held there safe from removal by more percolating water. The tremendous amount of surface in the clay fraction of the soil is the place on which plant nutrients are adsorbed. Even the silt fraction may manifest no small amount of similar performances. It is through this phenomenon of adsorption on its surface that clay serves in holding the fertility elements against loss by leaching and yet in readiness for delivery to the plant roots by exchange mainly for hydrogen given off by the plant.

Perhaps in the simplest way we can picture the plants' getting nourishment from the soil largely as a business of barter or trade. The root, as a colloid with its extensive surface giving off carbon dioxide that provides hydrogen, is in intimate contact with the clay surface on which fertility elements, such as calcium, magnesium, potassium and others are adsorbed. The hydrogen from the root is a positively charged ion and is traded or exchanged for those of similar charge on the clay. These positive ions, or nutrient cations, on going into the root are synthesized into complexes and carried up into the plant to clear the way for others to follow. Thus, the plant gains nutrients, or "takes" the soil fertility from the clay to reduce the fertility supply there and in turn to increase the supply of hydrogen on the clay surface.

The performance may be illustrated by means of the accompanying diagrammatic sketch, chart 1 on page 62, in which the hydrogen is designated as moving from the root to the clay along the lower sides of the circles representing clay or humus colloids, while the calcium is moving from the clay to the root along the upper sides of the circles.

Since it is the hydrogen ion that represents acidity, or sourness to our taste, the increasing concentration of hydrogen on the clay means increasing soil acidity. Plant growth and its removal of fertility from the soil is then truly a process that makes the soils more acid, in turn, increasing the degree of exhaustion of the fertility store from the soil. For plants growing on the more acid soils, then, we may expect that they are running a manufacturing business with an output of products with fuel value rather than a value because of their high concentration of protein or of body-building minerals. Acid soils are of low fertility and consequently of lower productivity. Their vegetation shifts toward that of mere fuel value.

If the plant obtains its nutrients through surface contact between the root and the clay because of the hydrogen as the root offering in exchange, the amount of nutrients so obtained from the clay of the soil will depend on three factors. The first of these is the total amount of root surface the plant has to offer per unit volume of soil. Densely rooted plants get more fertility than those only sparsely rooted. Bluegrass with 66 square inches of root surface per cubic inch of soil can take more by exchange than soybeans with only 2.5 square inches in the same soil volume. The second factor is the amount of clay surface, rather than surface of silt or sand the soil has, since little of fertility is adsorbed on these latter mineral separates of larger sizes. The third and important factor is the degree of saturation of the clay by the nutrients, rather than by hydrogen. Much clay in a soil means more chance for the roots to make contact with nutrient-carrying surfaces. Here is the reason why heavy clay soils are appreciated for the productivity, even if often hated because of their intractability. Ancient civilizations on sandy soils have not been long-lived. Those on clay soils have persisted through centuries. Regions of older civilizations today are on soils of high clay content because only such a soil would retain its productivity through those long periods of cultivation. But by far most effective in raising the productivity of the clay is the degree to which it is saturated with fertility rather than with hydrogen or with infertility. This variation in degree of saturation is the hidden condition that is not recognized and in it many of our soil problems originate. By it much can be done for their solution.

Though Missouri soils tend to be high in acidity, it is well we remember that if a soil can hold much hydrogen to make it acid, there is a large capacity to hold other positive ions that can be fertility, when we once decide to apply it. Missouri's location has been responsible for giving her soils much clay as well as a high degree of acidity in that clay. Her soils have the capacity to hold fertility. This fact offers hope for long continued use if our management puts the fertility into those soils.

Another glimpse at the diagrammatic sketch of chart 1 suggests that the clay and the silt minerals undergo interactions of exchange, much in the same manner as is true for the clay and the plant root. This is one of the fundamental facts pointing to the possible use of the reserve fertility in the mineral crystal of the soil. The clay plays an important part as intermediate agent between the plant root and the mineral crystal. The plant can get nutrients by direct root contact with the mineral crystal, but plant growth by this means has been experimentally found to be at a lower rate than when an acid clay serves as interceder between the plant and the mineral crystal. We can visualize the clay (a) as an acceptor of hydrogen from the plant root, (b) as a conveyer of it to the mineral in concentrations of significance for releasing the nutrients in the mineral by this acid effect, and then (c) as the deliverer in return to the plant of the nutrient set free from the mineral. The acidity in the clay is in reality of service rather than of detriment. Because the clay supply of fertility is rapidly exhausted, it is in the stock or minerals, particularly those of silt-size particles in the soil, that the reserve fertility and future productivity of our soils must be found.

The humus of the soil has always been recognized as of great help in soil productivity. It plays a dual role, it serves as does clay, because of its tremendous adsorbing capacity. Like the clay, it can take on hydrogen in trade for adsorbed nutrients. It, too, can exchange its hydrogen for nutrients in the mineral crystal. In addition, however, it decays rapidly and gives up for plant use the plant nutrients of which it was originally constructed by plant-growth processes. Clay does little of decay for nutrient release.

Humus has been playing the major part in productivity, because it does so in these two ways of conveying as well as providing nutrients of fertility value. Exhaustion of the humus is pulling down the fertility levels faster than anticipated. Humus is from five to ten times as effective as clay in the activities of exchanging nutrient cations. Its depletion from our soils is lessening the soil's effectiveness in production and bringing the problems of soil fertility rapidly to the forefront.

HOW LONG DO EFFECTS FROM FERTILIZER LAST?

Just how many successive crops on the land are benefitted by a single application of fertilizer is a question that has often been raised. It has seldom been answered except by estimates, consensus, and other answers that are only approached to an exact figure.

In Great Britain, for example, it is common agreement among landlord and tenant to consider the effect of a fertilizer as lasting for three years. Accordingly, if the tenant moves off a farm before three years have elapsed after applying the fertilizer, the landlord reimburses him accordingly as two-thirds or one-third of the fertilizer value is still left in the soil.

In our evaluation of returns from fertilizers we have been prone to calculate them for a single fertilizer treatment by charging the entire cost of the applied fertilizer against the first crop following its application. If the increase in the crop, figured at its current prices, pays only for the fertilizer we are prone to say that the fertilizer paid no profit, or it failed to pay for the labor of distributing it. Has all of the possible value in the fertilizer been collected in so short a time? Haven't we been asking too much, or too speedy, return when we expect the first crop and only one crop following the application to be increased enough in its yield to pay for the fertilizer?

When we use only the increases in the weights of the crops during succeeding years from the soil once fertilized as they are larger than the crop weights from soils not treated, we find that the effect of the fertilizer is not carried over for many years. However, when additional criteria are brought in, such as the feeding quality or the discriminating selection by animals, then the testimony of the dumb beasts points out that we have not appreciated the long-lasting effects that soil treatments like fertilizers are giving us. We are getting annual returns on the fertilizer investment for many successive years when we turn to animal choice or selection for evidence.

For just how many successive years cattle, for example, can recognize the effects on the hay from fertilizer put on the surface as a meadow treatment was demonstrated as a final distinguishing figure this past year by a couple hundred head of them on the farm of E.M. Poirot, near Golden City, Lawrence County, Missouri. As a master farmer of Missouri, Mr. Poirot has been a close observer of the results from soil and plant differences. His farm is on the prairie area of southwest Missouri with Cherokee and Gerald silt loams as the soil types. Much of his farm is still in the virgin prairie state. The surface soil is a fine silt loam, but is underlain by a tight clay, sometimes called "hardpan," none too far below the plow depth. It is highly in need of lime and responds readily to most any soil treatment as fertilizer.

It is on these soil types that Mr. Poirot has been carrying on many tests as a cooperator with the Agricultural Experiment Station of Missouri. His virgin soil has been an excellent, undisturbed base on which to try different soil treatments. With the larger share of his farm under the plow, Mr. Poirot's observations on the growing crops, their yields, and their feeding values, have supplied the Station with much valuable data. His close scrutiny of a good-sized herd of cattle, with everything under careful notes and records, has also been the source of many suggestions as to the fertility problems of his soils.

During the past 20 years, Mr. Poirot's trials of soil treatments and his attempts to solve the problems of soil fertility have run an extensive course. He has used limestone in lesser amounts drilled like a fertilizer for calcium's sake more frequently than larger amounts for the purpose of removing soil acidity. His demonstrations of growing clovers with only fractional neutralization of soil acidity were among the pioneer efforts in getting a better understanding of the function of liming soils. His farm trials with artificial manure, made directly from the straw pile as the straw was blown out by the threshing machine through which he also delivered the nitrogen reagent for making it, attracted wide attention. It was during the last nine years that he has been gathering observations of choices by cattle among the haystacks to give an answer to the question as to how long the effects of fertilizers can last.

It was early in the spring of 1936 that he undertook to topdress some virgin prairie meadow by drilling fertilizers of various kinds of surface applications. In this attempt to improve the natural grass crop without reseeding or plowing, a variety of different fertilizers were applied, each at increasing rates up to a maximum of 300 pounds per acre. One combination of two different fertilizers was applied each at increasing rates up to a maximum of 300 pounds per acre. One combination of two treatments were put across one end of a field, covering about four of the total of 100 acres. In order to supply calcium, the nitrogen was used as calcium cyanamid.

During the summer of that year, 1936, observations were made on the changes in the flora. Quadrant areas were harvested. Samples were taken for chemical analyses and other attempts made to evaluate the crop differences by the customary measures. Differences, though small, were detected and recorded.

In the following September the prairie grass was cut for hay. That from the area of slightly more than four acres given soil treatment the preceding spring was part of an area of 25 acres that went into the first stack. Three additional stacks of hay from the area without soil treatment were made of 25 acres each as the balance of the field.

It was October when the cattle were turned into this 100 acre field from the end opposite that where the soil treatments were applied, because of the location of the water and salt. Quite as a surprise, the cattle soon were all concentrated about the one stack consisting in part of the hay from the fertilized soil area. The other three stacks were disregarded as daily the cattle went by them in going back and forth from the water and salt to the chosen feed. This choice stack of hay was consumed completely by more than 200 head of cattle before the other three stacks of hay from soil without treatment were taken.

After 1936, no additional fertilizer applications were made. But each year since then the hay has been made regularly. It was cured and stacked in the same regular manner as four stacks of 25 acres each. The cattle have been regularly put into the field in the fall season to graze the remaining grass and to eat the hay. Each year through 1943, the cattle have taken first this one stack into which there was mixed the hay from the soil given fertilizers in 1936 along with the hay from soil given no such treatment.

During the eighth demonstration, in 1943, of the cattle choice which proved that the effects of the fertilizer had lasted for eight years or eight successive crops of hay, the discrimination by the cattle was still particularly keen. While making the hay, the stack bottom was laid down of dimensions not large enough to include all of the hay from 25 acres. Consequently, after all the hay from the treated soil was already in the stack, this stack was extended by adding at one end more hay from the soil without treatment. When the cattle were turned in from the end of the field given soil treatment to take the hay, they again chose this stack in preference. But instead of taking the entire stack, they consumed only that part with the mixture of fertilized hay, cutting the stack in two so as to leave the end made of hay from soil with no treatment. When this remnant was all that was left, then the cattle distributed themselves to the other stacks. Here was definite evidence that in the hay crop, even eight years after the fertilizer was applied, there were still evident to the cattle some qualities in the hay produced in consequence of the fertilizer put on the soil. The cattle were still recognizing some quality to provoke such discrimination.

The repetition of the demonstration in 1944, however, proved that the cattle were no longer able to recognize difference in the hays. When they were turned in this year, as in 1943, they consumed the first and second stacks nearest the water supply at about the same rate. There was no discrimination shown between the hays after nine years. However, when the cattle were grazing on this meadow in the fall of 1944, they invariably stayed on that part of the field where the fertilizers were originally applied.

This would suggest that the effect of the fertilizer was still revealing itself through the green grass, but was not detectable by the animals in the dried grass or hay.

Here is then a report by the cattle that they prefer the grazing and likewise the hay from fertilized soil. They give this report with keen discriminations. They also tell us that for eight and possibly nine years after fertilizers are applied there are returns to be had in terms of preference by the cattle, on the investment in this addition of soil fertility. When cattle indicate such choices as these were for their better feed consumption and therefore for better gains, one should not have doubts as to whether fertilizers pay dividends over more than one year. The animal's judgments on the Poirot farms are saying that fertilizer effects have been lasting for eight years, at least.

ORGANIC MATTER—THE LIFE OF THE SOIL

The cornbelt, or at least the more intensive agricultural belt of the United States that superimposes so snugly over the glaciated area and provides the maximum concentration, or even the maximum total, of agricultural buying activity in this country, derives this power from the organic matter, or from the life of the soil. Dark color of the soil, whether in the cornbelt of the United States, the wheat belt of Russia, or the Pampas of Argentina comes from the organic matter residue of past plant generations. This residue is the basis of production of those foodstuffs for plants, animals and man, that make for a well-rounded and highly diversified agriculture. The dwindling supply of this organic matter heritage of the ages past is reducing the flame of production to a point, not only of economic irregularities, but even to the lack of providing enough vegetative cover to hold back the waters that carry the very soil itself away. It is high time that we look nationally into the "life" of the soil if we are to sustain our own lives, so directly dependent upon it, to say nothing of passing some remnants of a natural soil heritage on to those who follow.

It is an age-old truth that manure is good for the soil. We shall but remind ourselves that organic matter supplied through the old manuring practice makes for a delightful physical condition that every farmer recognizes as a granular soil structure to encourage water intake, deep root penetration, resistance to drought, ease of cultivation over a wide range of conditions, and many properties of these soils readily recognized as being well loaded with humus, or organic matter. Along with these desirable physical properties goes the capacity to deliver good grain harvests and nutritious forage yields. Such deliveries are not the physical but rather the chemical performances. They are the unseen behaviors within the soil, and determine those visible or physical properties which change when those of chemical nature change. Crop production is promised on these chemical changes. The declining production level is paralleled by lowered chemical stocks or lessened rates of chemical transformation of them.

We have been prone to believe that high crop production is caused by the high level of organic matter. Because two things are performing together is no proof that one is the cause of the other. Both may have a common cause. Because two women are jostling each other in the doorway of a store, can we say that one or the other is the cause of it? It may be the bargain sale on the inside that makes both behave in that low order. Similarly, we should be raising the question whether high crop yields and high organic matter of the soil are not both the result of the same causes, and whether our declining crop yields and declining stores of soil organic matter are not resulting from the declining supply of fertility elements within the soil.

Fallacious reasoning of this type is to be expected unless our knowledge through research, and other forms of mental industriousness bring us to the common cause: the recent increased study of soil. That has brought this subject into the list as about the last of our newer sciences, and is responsible for our better appreciation of the 3% of the soil that is made up of organic matter. We recognize now that it is within this small soil fraction that the real activity in plant nutrient manufacture take place during the growing season. Its level of crop output is not determined by the fact that it is just simply carbonaceous organic matter, or is combustible, but that it is organic matter produced under soil conditions offering liberal supplies of calcium, phosphorus, magnesium, potassium and the other nutrients drawn from the soil rather than from the atmosphere. Organic matter within the soil, but initially produced on a soil that had a high stock of calcium, phosphorus, potassium and other nutrients so that it is now delivering by decay, liberal fertility allotments, is then the sound reasoning for high crop production on dark soils.

If then liberal stores of fertility delivered by the soil and generous stocks of organic matter into which much of this fertility has been put in organic combination within the soil are causal to high crop yields, the crop production picture is still much confused or mixed as the organic matter itself is a mixture. Let us attempt to elucidate this crop and soil interrelation by reducing it to simplicity. Let us ask and answer two questions; namely, of what is a plant made, and what or how can the soil contribute the making?

The soil's contribution to a plant is represented by the ash or the noncombustible part. This amounts, generally, to less than 10%. Consequently, 90% of the plant is made up of carbon, hydrogen, oxygen and nitrogen coming from the air and water, or is contributed by the weather. The magnitude of this contribution by the weather is, however, no excuse for blaming the weather for the 90% or more of our poor farming. The 10% contributed by the soil, and neglected by many of us farmers, is no less of importance than you would give to the small heart weighing but about seven-ninths pound in a body that weighs 150 pounds.

The soil's contribution involves ten elements. Four of these; namely, boron, manganese, copper and zinc, are used so scantily by plants and were mixed so abundantly into our soils by the glacier plowing down from the north over all kinds of rocks, that we are seemingly suffering no shortage in the cornbelt of these four minor elements. As for the remaining six elements, which include calcium, phosphorus, potassium, magnesium, sulfur and iron, the matter dare not be dismissed so light heartedly for all of them. With a very small need by plants for iron, which is less than ten pounds per 100 bushel corn crop when there are over 90,000 pounds of it in an acre of plowed soil, we are not much concerned about an iron shortage. With sulfur going into the air from burning coal and washed back to the soil by rain at rates as high as 25 pounds per acre per year while crops of corn of yields as above use but 12 to 16 pounds, there is here no cause for grave concern. Magnesium has been announcing its shortage only in some particular soils and for some particular crops, but not significantly so in the cornbelt. No harm comes from its addition, however, to the soil in the form of dolomitic limestone in place of the pure calcium stone. The potassium situation is not solved by natural contributions from the rain, as is true for sulfur. But, the soil is well loaded with mineral potassium, if only it can be gotten out of the original rock into a form suitable for plant use. Fortunately, potassium is brought back to the soil in the plant roughages, as straws and stovers, and but relatively little is removed in the crops, and the animals.

Thus the plant nutrient contributions by the soil are ample in eight of the ten items, purely because of large soil store and little consumption by the plants. But as for the remaining two, namely, calcium and phosphorus, these are foremost soil deficiencies. Of these, calcium deficiency is our national problem number one.

What about nitrogen, you may be about to inquire? In a soil conservation type of farming, to which we are now moving, the use of legumes will solve the nitrogen problem, but only when calcium and phosphorus are provided to make those legumes effective as nitrogen fixers. Many of our newer legumes have been deceiving us to the point of posing as soil improvers, when the calcium and phosphorus offerings by the soil were too low to let them serve in this capacity.

How does the soil deliver these nutrients as components of 10% of the plant structure, is the next question. Soil consists of sand, silt, clay and the organic matter fraction. The organic matter makes up less than 5% and of the remaining 95 parts, only about 10 to 12%, or less than one-eighth of the soil, is all that is of nutrient possibility, even if it could all be transformed to a soluble form that would render such service. Basically, 90% of the soil you own is of no plant food value. It is only a place where plant nutrients may be put. Remind yourself that this 10 to 12% of plant nutrient capacity has come to us after resisting the weather and the transformation into the soluble conditions for not only centuries, but for geological eras. Had it become soluble, it would have gone to the ocean long before we arrived on the scene. It is from this store with ability to hold out against weathering for ages that we ask our plants to satisfy their ash needs through contact during a short growing season.

The sand fraction, insoluble enough to come down through the ages intact as larger particles, consists of quartz and contains no plant nutrient contribution. The silt may be as much as 60 or 70% of the same quartz material and leaves only about one-third of this fraction bearing nutrient elements. But again here they are in the original rock form, too slowly soluble to be in reach of the plant. Even the clay minerals, more finely divided than the silt, contain little of plant nutrient within their mineral crystal structures. Fortunately, however, clay has the capacity, not possible for silt and sand but characteristic of humus, to have adsorbed on its surface much of nutrient significance and in a form readily exchangeable to the plant but not so readily leachable or lost by percolating water.

Thus, as soil minerals became slowly soluble, their nutrient cations may be caught up or adsorbed by the clay, held there against leaching, and exchanged to the plant for its offerings of hydrogen or acid. Here, then, is the active mineral fraction of the soil to serve along with the organic matter fraction as help to the nutrition of the plant. Nutrient delivery by a soil to the plant come in mainly from two sources, first, this adsorbed supply on the clay surface where it was collected by ages of slow mineral weathering and from which it can be quickly exhausted by plant roots, and second, from the breakdown of the past plant generations which give up their ash either to be adsorbed on the clay, or on the remaining organic matter humus, or to be taken by the plant, or to be lost by leaching.

For activities during the growing season, we need not look to the sandy nor to the silty part of the soil. On the clay there is reposing a small stock of exchangeable calcium, magnesium, potassium and others, but the whole of which will not move into the growing plant. The main store then to be delivered for crop production must come from the organic matter that was in plant forms once before. We are then rebuilding our crops today from the ashes of the previous dead crops. We are burning the dead past of the eons of time to make the plants of a single year. We are exhausting rapidly the greatest of our natural and national resources with little attempt to understand the decline, and less to restore in some measure that which has been consumed.

Organic matter becomes the life of the soil because with each opening of spring, the rising temperature starts the bacterial wrecking crews to work on this carbonaceous residue. In their struggle for energy and nutrients, they must tear apart the old compounds. They burn out the carbon which escapes as carbon dioxide and they respire the hydrogen and oxygen to become water just as you and I do with the carbohydrates we require. They consume some of the nitrogen, convert some into nitrates, and use some of the ash residues for growth. They discard much of the mineral fraction. While the carbon escapes and other elements remain, it serves to unbalance their diet in this energy providing component; the minerals are then available for the plants.

Here are the fires within the soil that give the plants the first items to start their growth. Here is the basic farm factory that must run first before the plant can catch the sunshine to run its carbohydrate, its protein, or its vitamin factories. Would you believe that 40 acres of corn land making better than 45 bushels to the acre are burning enough organic matter daily during the July season to run a 40 horsepower steam engine? Have you ever thought of your responsibility of feeding that wrecking crew of bacteria working for you so faithfully within the dark recesses of the soil? Viewed from this standpoint, organic matter readily provides the life of the soil. *Crop production then becomes as simple as the mere task of feeding the helpful bacteria within the soil, more than so much concern about fertilizing the crop above the soil.*

In reply to your query as to what we must feed the bacteria, may we counter with this query, how about plowing under a crop of red clover? Doesn't this increase the corn crop following? Doesn't the soil plow better and cultivate more nicely where legumes like red clover or alfalfa have grown and are turned under? Feeding these soil bacteria on a red clover diet means better farming than feeding them on a wheat straw or cornstalk diet, just as red clover and alfalfa feeding of animals is better livestock business than keeping them on straw and corn stover. Soil bacteria, like animals, do better on legume hays that were made before the crop was too mature. Just as you won't expect to raise young animals on straws and stovers and expect work or profit from them, so you must not expect to raise a good crop of soil bacteria and get them to run the soil factory with profit, unless they are fed a ration of the red clover or alfalfa standard.

Whatever is good for the bacterial crop within the soil is also good for the grain or grass crop above the soil. When soil conditions produce good clover, or good alfalfa, they grow good bacteria to run the soil nutrient factory. Conversely, if the soil won't grow red clover or alfalfa, the bacteria are not doing much good in the soil. Soil conditions that grow red clover represent a good supply of calcium, of phosphorus, of magnesium, of potassium, and other ash constituents. This means good protein by legumes through nitrogen fixation from the atmosphere. This means a balanced bacterial diet when plowed under so far as nitrogen is concerned. The bacterial ration must be balanced by nitrogen just as the animal ration must have the carbohydrate, corn, balanced with a protein such as tankage or cottonseed meal. If it not balanced, the bacteria may balance it for themselves to the detriment of the crop on the soil. Wheat straw plowed under, soybean straw, sorghum roots, and other similar carbonaceous matters, fed to the soil bacteria will force them to search the soil for soluble nitrogen. They will reduce this supply to the point where wheat after wheat cut by a combine may fail for want of soluble nitrogen. Strawy manures behave similarly. But if rotted, much of this surplus carbon is respired away so as to narrow the carbon-nitrogen ratio and eliminate the danger to the following crop. Bacteria in the rotting manure give off carbon as heat and make the manure a balanced diet for the bacteria in the soil. Red clover plowed under gives excess of nitrogen compared to carbon in the bacterial diet, and they throw off some soluble nitrogen. This nitrogen is then available for the crop. Young sweet clover serves similarly but woody mature sweet clover is a ration that the bacteria would put into the same class as straw. Bacterial diets must have a 25 to 1 nutritive ratio or less, in terms of carbon to nitrogen if they are to deliver nitrogen. These are requisites for soil bacterial rations just as animals demand their specific nutritive ratios in feeds if such are to be profitable.

Red clover and other legumes are of this proper nutritive ratio. Bacteria thrive on them, as evidenced by the case with which clover hay spoils in the field, or the rate at which bacteria can grow on it to consume it. Making timothy hay is no problem because it is a poor feed even for the bacteria. Boiled potatoes are consumed slowly by the hungry boy in contrast to their rate of disappearance when mashed and immersed in beefsteak gravy. Is it any wonder that young animals don't do well on timothy hay that won't nourish soil bacteria effectively? Have you ever carried these crop composition differences back to the differences in soil fertility requirement to grow these crops?

The nitrogen shortage in our soils is one of the main reasons why organic matter has accumulated to some extent in our soils, in place of being consumed by the bacteria. It has been like so much bread, awaiting meat slices as nitrogen, or mustard condiment as minerals to enable the bacteria to live on it. This nitrogen problem can be solved by the growing of legumes, but these will not draw on the air for their nitrogen unless the soil offers the minerals.

Legumes are balanced bacterial diet partly because of balanced nitrogen and carbon in the plant composition, but also because of balanced mineral contents found there. Too little have we thought that soil bacteria require calcium, phosphorus, magnesium, potassium and other soil items. Liming many of our soils has been feeding the bacteria calcium to enable them to decompose the organic matter and set free all the other nutrients tied up in it. Soils so low in calcium have been short in this item in the bacterial ration and in the crop ration, too. Corn grows better after liming because the bacterial crop is fed calcium and because some extra calcium is provided for the corn crop. Legume nodules are most numerous where the roots meet decaying organic matter. Is this because it offers nitrogen? Scarcely for a legume that uses nitrogen from the air, but more probably because the decayed organic matter gives calcium which is so deficient in most soils for effective nodule production and nitrogen fixation. "Soil acidity is not so severe on soils rich in humus," you say. This rules out the degree of acidity as determining the damage, but emphasizes the calcium delivered by soil acidity.

These illustrations are sufficient to point out that with high levels of calcium and phosphorus in the soil, there can be good growth of legumes to capture nitrogen from the air, to unite it with carbon, to go back to the soil as organic matter or a well-balanced diet for the bacteria, and to be decomposed with liberation of its nitrogen and mineral for rapid production of another crop, and then to repeat the process. Thus, our black soils, because they contained calcium and phosphorus in liberal amounts, are oriented to encourage proteinaceous legumes to build up nitrogen in the soil. This liberal level of nitrogen serves to hold carbon as a great deposit of humus. This great deposit of mineral-rich and nitrogen-rich humus as a balanced diet is set to give us the speedy breakdown of this organic matter heritage that we have now exhausted. Black soils are rich in activity in growing organic matter above the soil; in activity in decaying the organic matter; and in the maintaining of a liberal stock of humus residue to keep them of black color.

The mobile lime, or calcium stock, and the high adsorptive capacity in our cornbelt soils have been the basis for proteinaceous production not commonly appreciated. Neglect to restore the calcium stock in the soil, as the main key in protein manufacture by the plants; to provide phosphorus as a constituent of protein; and to use these through the legumes to maintain our soil organic matter at a high level and at a high degree of activity, is rapidly moving us from proteinaceous crops to carbonaceous crops. We are moving ourselves southward in equivalent, where the soils depleted by nature of calcium, of phosphorus, and of absorbing capacity are still offering potassium to the carbonaceous crops like cotton, sugar cane, or pine trees, but are doing very little to provide calcium and phosphorus to grow life sustaining, proteinaceous products. Do we want to reduce cornbelt farming to the troublesome conditions of the South? Do we want livestock production on the southern mule level, the animal which is imported at maturity from the North. and whose failing hope for posterity fails to reveal the trouble in reproduction by animals fed on soils of southern fertility levels?

We need to look to the fertility of the soil to maintain the organic matter that runs our soil factory for the plant nutrient production. We are now at the crossroad. We have ridden the toboggan down the hill of fertility decline from the accumulated heritage of the past, and must either push back up on the hill through organic matter restoration, or trudge along slowly on nature's level of equilibrium dragging our toboggans with us. Organic matter decay within the soil to liberate its fertility will still be the foolproof method of providing nutrients for the crops. Commercial fertility can and must be used, but will be most effective as it is held back in organic matter combination to be released slowly rather than be allowed to leach from the soil as salts. Calcium for the legumes to encourage nitrogen fixation, and phosphorus as a supplement that is most effective only on calcium-laden soils are the two main requisites—possibly to be supplemented later by others—that must be used to gear our land into legume production. Much of this legume organic matter must get into the soil. Not all legumes will serve in organic matter restoration. We have slipped downward from alfalfa to red clover, to sweet clover to soybeans, and to lespedeza on side-stepping the soil's demand for calcium and phosphorus. We have accepted lower fertility level crops. We have taken them into the legume crop category of nitrogen fixation without realizing some wide differences in composition between them. At the same time we are accepting lower quality in nitrogen fixation and feed production and doing little or nothing for organic matter return to the soil. We must go back to the higher fertility idea represented in alfalfa and red clover. With the good soil conditions demanded by these as ideals for organic matter restoration in the soil, and with our contributions of lime, phosphorus and other soil fertility decline will be halted, or even reversed, as we restore or maintain our soil organic matter, and really "put life" into the soil.

MAKING ORGANIC MATTER
EFFECTIVE IN THE SOIL

Organic matter has long been empirically recognized as the key to soil productivity, but an understanding of how it serves as such is not of long standing nor of common knowledge. Like many other biochemical behaviors and life processes, its effects are recognized; but its modus operandi is not known. It may be fruitful in practice to understand some of the simpler principles around which the behavior of organic matter in the soil may be classified.

Perhaps it may be helpful at the outset to establish our philosophy regarding organic matter in the soil. Because it is so significant in soil productivity, and if it is true that the more organic matter a soil contains the more productive it is, it seems logical then that we should busy ourselves in raising the organic matter content in the lighter colored soils by intensive crop growth on them to be turned under for green manure. Chemical analysis of soils which are seemingly low in carbon and nitrogen leads students to deplore the productivity of a soil in the light of such determinations. Seemingly then our philosophy might lead us to say that we should try to increase the organic matter in every soil from any level found there.

This viewpoint is to be expected from the general feeling always going with big sums, large concentrations, or other big things, that such must be valuable merely because of magnitude. Our general classification of darker soils into the higher category of productivity is generally correct, but does not deny even good productivity to many soils lower in total organic matter. It is essential to recognize that organic matter should be considered as the one condition of concern about the soil for productivity. This can be taken from empirical thinking when we remember that manuring has come to us from ages of farming experience. Maintaining the organic matter would certainly be a good philosophy even if it came by rule of thumb and if it had no further bases. We do not want our fertility to decline. The decreasing organic matter is the first index of a slipping productivity even if we don't appreciate declining crop yields. Maintaining the organic matter at the level we found it certainly would be good philosophy, should we find it economically unsound to push a soil's organic matter content significantly upward.

Whether we should undertake a national campaign to increase organic matter extensively and intensively might not be doubted by the ambitious of the crusader-minded conservationists on recognizing the extent of our abandoned lands in the humid area. However, the attempts to shift natural equilibria usually result in disaster for the self-appointed shifters. Those who would buy poor land to rebuild it by the organic matter injection method usually report about as much success as the delightful maiden who marries the liquor addict in the hope of curing him of his habit. Fatalities in the former are not as numerous in total as in the latter, mainly because the attempters have not been as plentiful. On a percentage basis they are perhaps higher.

Thus we certainly are not given to the philosophy of hoarding the organic matter in the soil. Unless it decomposes to liberate its nutrients it cannot contribute these to the growing crop. Organic matter in the soil, like money, is not of value *per se* but in its turnover, or change. Can we believe otherwise, the sudden increasing of the soil organic matter? Doubtless not, for sudden changes in nature usually meet much opposition. Experience of trying to measure effects on soils because of immense organic matter additions puts their benefits so low as to eliminate much expectation from them. Perhaps then our proper philosophy would suggest that we need first, to guard jealously, but not hoard, the soil organic matter; second, we would scarcely dump large quantities of crop residues into the soil and expect them to build up in semi-permanent form the organic matter stock in the soil; and then third, as a middle course we should accept as proper philosophy the attempt to maintain, at least, the organic matter level, but aim to put back into the soil the largest share of organic matter produced on the soil commensurate with more reasonable use of land for crop and animal production and human sustenance on it.

In the light of such a philosophy of good stewardship toward the fleeting productivity property of the soil, let us next become acquainted with the functions and services by the organic matter. Immediately the physical effects by organic matter on the soil properties will flash through your mind as it recounts, (a) delightful granular structure, (b) improved workability, (c) better water reception and consequently less runoff and erosion, (d) improved drainage, (e) darker color, and a host of other physical properties by which the plowed-out sod land impresses you in contrast to the soil in continuous arable farming. More significant than these visible properties are the invisible chemical properties and biochemical performances that control those physical. Granulation is not the result of the presence but of the decay of organic matter. Soil structure is modified by the secondary products resulting because the organic matter has undergone extensive bacterial attack and change. Organic matter is not a stable situation, but a changing condition that is valuable through this fact of change. Virgin soil is impressive because the maximum of decay of organic matter within and of growth upon the soil are occurring annually to give the highest possible level of changing rather than of static organic matter effects.

Perhaps the most significant, but not recognized, aspect of the organic matter change is its synchronization with the growing season. Changes of soil organic matter increase their speed with the rising and decrease with the falling temperature. Life processes of bacteria in tune with the growing season are liberating plant nutrients from past plant generations for those of the present at rates commensurate with the temperature, and, therefore, with the growth of the crop (see figure 1). There is thus no large supply of soluble fertility dumped with the seed, but all through the soil there are increasing amounts of nutrient wastes coming from bacterial activities within the soil as the increasing and extending root system is there to consume them. Commercial fertilizers may be designed for resistance to soil adsorption, or be equipped with timing bombs for delayed release; but it seems doubtful whether they can escape leaching and whether they will ever deliver nutrients as efficiently as nature's own system of

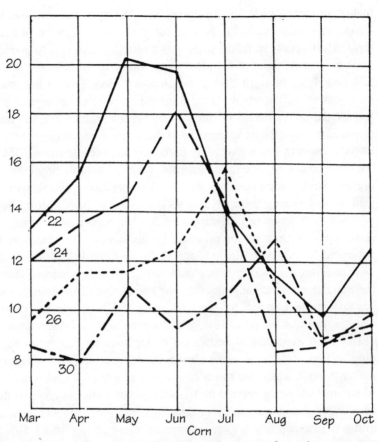

Figure 1. Nitrate accumulation as it reflects organic matter decay in agreement with the season and decline with time.

nutrient delivery from an ample stock of organic matter decaying within the soil. Organic matter will still be the premier, foolproof fertilizer even when scientific efforts at fertilizer compounding or even solution culture and soilless plant growth rise to a high degree of perfection.

The general use of legumes as contributors of organic matter to the soil gives emphasis to their offering of nitrogen to be the almost complete disregard of other contributions. It is not the aim here to discredit the use of organic matter for nitrogen sake. Rather the nitrogen deserves more emphasis since it is only by adding nitrogen to the soils that increasing amounts of organic matter can be held there. Nitrogen additions are the means of increased reserves of soil organic matter. Some chemical soil analyses are in point here. A soil under fallow with single annual plowing, but no mechanical losses or contributions, lost 115 pounds of nitrogen from 1918-1932. Under rye cropping with this carbonaceous matter all turned under in the spring and fallowed, the loss was 640 pounds during a similar calendar period but three years longer. Where cowpeas replaced this post-rye fallow there was a nitrogen gain of 120 pounds. Where a good crop of 2.5 tons of red clover was turned under annually this soil gained 324 pounds of nitrogen during the shorter period. From such results we can reemphasize the already accepted principle of building up organic matter in the soil by the incorporation of nitrogenous organic matter. We can also raise the question whether we have appreciated the nitrogen and organic matter depleting effect of turning under extremely carbonaceous organic matter. If this latter is the truth, as our Missouri data indicate, we need only to emphasize legumes as organic matter additions, but should also emphasize the danger of nitrogen depletion by turning under organic residues short of highly nitrogenous nature.

Another function of organic matter that has gone almost unrecognized is its contribution or release of minerals by its decay. This fact has been overshadowed by (a) emphasis on nitrogen, (b) seeming insignificance of the small amount of nutrient minerals in plants, and (c) belief in nutrient contribution by the mineral soil fraction. Now that our knowledge of the nitrogen significance has become understood, we have taken time to look into the contributions as calcium, magnesium, potassium, and others by organic matter decay.

Have you ever studied nodulation of legumes by digging along their roots only to find nodules of sweet clover, for instance, in every clump of organic matter through which the sweet clover root penetrated? Can this occur because of the nitrogen being offered by this clump when this exaggerated development of nitrogen-fixing machinery is right in the organic matter? Is it shortage of nitrogen that handicaps sweet clover on most of our soils? Can't it be the liberation of calcium so necessary on most soils for this crop and of which the content in plant ash is high, namely, as much as 35% in case of wood, for example, compared to 5% for potassium? Calcium is further suggested as of causal significance when you recall the common observation of farmers who have often reported that "soil acidity isn't so dangerous on the more fertile soil," how do they measure fertile soils except by organic matter content? We need to look to organic matter as the maintainer of our cycle of mineral turnover just as we look to it as a source of the nitrogen cycle, even though the amounts of minerals so handled may be small in comparison to the nitrogen involved annually.

In pursuit of the possibility that soil organic matter contributes other nutrients beside nitrogen, humus extracts were made of the soil from six different plots from Sanborn Field at the Missouri Experiment Station. These were analyzed for their contents of calcium and phosphorus as well as of nitrogen and carbon with the results given in table 1.

If we should assume that this humus is the whole supply of the nitrogen for a corn crop, it is interesting to calculate the corn possibilities in terms of the calcium and phosphorus released by this humus breakdown. The 3.21% of humus, or 64,200 pounds per acre, in the continuous corn plot with no manure, contained 3.45% nitrogen, or 2,256 pounds per acre. With 20 bushels of corn, as the average yield on this plot, that would require 30 pounds of nitrogen annually, which release of nitrogen represents breakdown of 1.31% of the total humus in the soil. The corresponding amounts of calcium and phosphorus that would be released along with the nitrogen are 17.9 and 3.79 pounds, respectively. This is the calcium needed for a corn yield of fifty bushels and the phos-

TABLE 1

Analysis of Extracted Humus from Some of Sanborn Field Plots.

Crop	Treatment	Soil content %	Calcium %	Phosphorus %	Nitrogen %	Carbon %	C/N
Corn	Manure	3.280	2.24	.710	7.40	14.15	1.91
	No manure	3.218	2.12	.448	3.45	8.76	2.54
Timothy	Manure	4.712	1.71	.648	7.09	14.11	1.99
	No manure	3.314	1.71	.842	5.52	16.51	2.99
Rotation	Manure	3.958	2.74	.724	5.34	18.09	3.38
	No manure	3.322	1.56	.808	4.99	17.07	3.42

phorus for 25 bushels, both of which figures are yields above that being produced on the plot. If this assumption holds true, then the organic matter, or humus, could supply not only the nitrogen, but the needs of the crop for calcium and phosphorus as well.

More emphasis to the mineral contribution by organic matter decay has come from the observation that the mineral fraction of the soil may contribute little or nothing for plant nutrition. Can it be possible that a soil of 95% or more mineral and 5% or less of organic matter can not be contributing mineral nutrients from this mineral mass? Doesn't soil weathering give soluble mineral materials? Experimental studies of the most finely divided mineral fraction suggest a negative answer that would consequently leave almost wholly to the organic matter fraction the responsibility of supplying the mineral nutrients.

The finer clay fraction, almost free of organic substances, was electrodialyzed to remove its exchangeable nutrients and then reloaded with known amounts of specific nutrients as, calcium, magnesium, and others. This made possible, by clay analysis and chemical additions, a complete balance sheet of the chemical situation at the outset of the plant growth or of the carbonic acid saturation to learn whether these periods of exposure of the clay to the plant roots or to carbon dioxide contributed any soluble situation at the close. Plant growth of six weeks, or carbonic acid treatment for the same period, failed to release calcium, magnesium, potassium or other nutrients of significant amounts from this finer crystalline mineral fraction of the soil. With no release from this finer soil fraction, surely the coarser silt and sand fractions, which are made up mainly of quartz which carries no nutrients, cannot be expected to give significant mineral, nutritional aids. We are then forced to emphasize organic matter not only as a source of nitrogen, but also of mineral nutrients during its decay for most of the plants during the growing season, because the mineral soil part is not contributing them.

Plant growth is thus in no small measure a case of rebuilding from the nutrient constituents of preceding plant generations. Recognition of such a principle will be helpful in the light of its recognized truthfulness, in some phases of life, as, for example, education, where we do well to get but a fraction of what has gone before. Can we not look to the composition, then, of the plants growing dominantly on the soil as an index to the offerings in nitrogen and minerals by the organic matter decaying within the soil? The plant composition in terms of total fertility will indicate the level of fertility offerings by the soil. Will this not give us a clue as to the level of nutrients, making the cycle, (a) into crops, (b) back to the soil, and (c) into crops again with some loss in the circuit? Perhaps, when we remind ourselves that bacterial life within the soil is responsible for the transformations, and when we look to these bacterial activities as possible distortions in the cycle of nutrients or even sidetrackers of it, this organic matter-fertility performance may become more simple.

In the bacterial life processes, energy is not drawn from sunlight as is true for plants. Soil microorganisms use the chemical energy set free when the carbonaceous compounds, the synthesis of which captured the sun's energy, are broken down to the simpler forms again of carbon dioxide and water. This is the source of energy for their activities that dominates their behavior as indicated by the carbon dioxide coming from the soil to the atmosphere. No less significant for the bacteria are the minerals and nitrogen that are synthesized into bacterial growth, or body construction, and again released to remain within the soil and crop cycle subject to some leaching loss, rather than to escape as gases to the atmosphere.

Should we disregard the carbonaceous phase that comes from water and carbon dioxide of the air and escapes from the soil to that place again by organic matter decay—and which is also outside of the limiting elements in plant production—we can then concern ourselves with the effectiveness of organic matter within the soil according as its supply of nitrogen and mineral nutrient is at a high level in this cycle of organic matter turnover.

Nitrogen and minerals in the organic matter are in turn dependent on the composition of the plants that produced it, and their composition in turn on the nitrogen and minerals offered by the soil on which they grew. Thus we come back, in the ultimate, to the problem of soil fertility in terms of the nitrogen and minerals at the crop disposal as they determine the effectiveness of the organic matter resulting there from within the soil. Thus the cycle of soil fertility going into the crop and from the crop into the soil organic matter is one whose magnitude is determined not so much by the carbon, but by the level of minerals and nitrogen in the cycle of fertility turnover.

On the basis of such reasoning we may expect that bacterial activity will also fit into this process and be higher when minerals and nitrogen in the usable form are at a high level. Extensive studies of organic matter decay in terms of nitrogen transformation and other chemical changes subscribe to the truth of this hypothesis. They show that the bacterial ration, like the plant ration, demands a balance in a liberal level of fertility, only to give unfortunate crop effects unless this bacterial ration is properly balanced. Bacterial behavior comes first and crop behavior comes second.

Feeding soil bacteria is no different in principle than feeding higher forms of life where the nutritive ratio of carbohydrate to protein in relation to nitrogen in the green manure or dry matter turned under, doesn't stunt the bacteria as much as such an irregular nutritive ratio does a hog. Bacteria will of their own accord, remedy the shortage of organic nitrogen in the compound by using mineral soil nitrogen—the only nitrogen of plant service. When compelled to do so they reduce this mineral nitrogen supply far beyond a soil saturation degree needed by the plant. Turning under strawy manure is disastrous for these reasons. Effective manure use means a balanced carbon-nitrogen condition so far as bacterial diet is concerned. Here again the use of a bacterial performance to balance it before putting it into the soil, or after the manure has gone through its heat in the pile, is better procedure than the use of fresh excessively strawey manure. The heating of manure is the release of the excess carbon while the nitrogen and minerals are retained and synthesized into complex organic compounds. When these compounds are simplified by decay they will release minerals and nitrogen relatively as the carbon escapes into the air.

Much of the carbonaceous plant material may thus be dangerous organic matter to deplete soluble nitrogen to the detriment of a crop on the land, or to lower the soil fertility as suggested by data for rye previously cited. This danger may be remedied by the use of commercial forms of nitrogen with it. Such use of nitrogenous salts means its rapid synthesis into organic complexes rather than its extensive loss by leaching. After its synthesis it is then subject to slow release in accordance with growth conditions of the crop needing it and the bacterial activity setting it free.

The production of artificial manure by adding commercial nitrogen to our straws, cornstalks and similar carbonaceous wastes is a means of deriving soil benefit from organic matter now wasted or burned, and of giving it greater effectiveness within the soil, to say nothing of it serving as a means of using nitrogenous fertilizer salts more efficiently through delayed action in place of leaching losses as high as 50% in case of its direct application to the soil. More economical use of purchased nitrogen, together with increased effectiveness, and a higher level of organic matter in the soil will result when straws after combines, or cornstalks after pickers, will go into the bacterial soil manager in balance with commercial nitrogen. Artificial manure production hasn't been appreciated. It will be when its value in terms of effective organic matter is recognized.

As the shortage of nitrogen in the crop residue turned under makes organic matter less effective, so the shortage of potassium, phosphorus, and calcium may be disastrous through organic matter behavior. Some studies where sweet clover, which is rich in nitrogen and calcium, was turned under as green manure on a limed soil show that such may induce a potassium shortage for the crop in the soil. The corn plants are green, but short. They fall over on approaching maturity and give only chaffy ears if they produce at all. Even wheat following the sweet clover a year later is of lower yield than that on the soil without sweet clover used as green manure.

Analysis of the corn crop indicate potassium shortage and potassium addition as fertilizer are beneficial. Effective use of this sweet clover green manure in a two year rotation depends on fertilizing the soil with potassium salts and also possibly with phosphorus. Other cases of disastrous effect on corn by sweet clover point to the danger in organic matter through induced potassium deficiency in the bacterial ration because of shortage of exchangeable or available potassium in the soil on which the green manure was grown.

That calcium should be significant in the bacterial ration may seem farfetched, but bacteria, like animals, need carbon, nitrogen, minerals, and even possibly vitamins, or certainly some complex substances which we may call "growth factors" until they may be proved similar to vitamins. Studies of organic matter decay in Sanborn Field show that calcium may be deficient in the soil to the point of making organic matter less effective. Even the calcium supplied in ordinary superphosphate may be a significant mineral supplement in a bacterial ration. The gypsum in this fertilizer has possibly been effective through bacterial action in many instances when the entire credit has been going to the phosphorus as a nutrient for the plants.

The significance of such items as phosphorus and calcium in connection with organic matter behavior was given attention in further studies of soils from Sanborn Field for the changes in nitrogen and carbon contents as occasioned by fertilizer additions in field treatments. The data are assembled in table 2.

It is significant that the lowest total supply of organic matter, 1.94%, occurred in plot 2 where a complete mineral fertilizer, 750 pounds annually of nitrogen, phosphorus, and potassium carriers, was used. These nutritious supplemented carbonaceous parts in the bacterial ration aided the bacterial digestion of the organic matter to levels below those of any other plots. Adding nitrogen only, as sodium nitrate, plot 30, suggests that the bacterial ration lacked phosphorus, potassium, and calcium as gypsum, to give failure in lowering the organic matter to the low level of plot 2 getting these in addition to the nitrogen, or to that of plot 29, where a more acid condition relieved the

TABLE 2
Organic Matter Content and its Composition as Influenced by Fertilizer Treatments on Sanborn Field Soils.

Plot No.	Crop and Treatment	25 years				50 years				Change				
		O.M. %	Carbon %	N %	C/N	O.M. %	Carbon %	N %	C/N	O.M. %	Carbon %	N %	Ligain %	
2	Wheat-Fertilizer	1.94	1.13	.107	10.5	1.75	1.02	.100	10.3	-.19	-.11	-.01	39.5	
5	Wheat— Manure 6T, 25 yrs. Manure 3T, 25 yrs.	2.62	1.52	.140	10.8	2.18	1.27	.119	10.6	-.44	-.25	-.21	48.9	
29	Wheat— Manure 6T, 25 yrs. Ammonium sulfate, alone 25 yrs.	2.37	1.38	.145	9.5	1.84	1.07	.081	13.2	-.53	-.31	-.64	46.1	
30	Wheat— Manure 6T, 25 yrs. Sodium nitrate, alone 25 yrs.	2.77	1.61	.171	9.4	2.24	1.30	.094	13.8	-.33	-.31	-.77	50.5	
23	Timothy— No treatment	2.29	1.33	.141	9.4	2.49	1.45	.135	10.7	.20	.12	.06	40.5	
22	Timothy-Manure 6T	2.91	1.69	.177	9.5	3.51	2.04	.195	10.4	.60	.35	.18	48.8	

shortage in these items somewhat. Organic matter accumulates because of the deficiency of minerals in the organic matter and in the soil which makes a defective bacterial ration and to lower the rate of, or to prohibit, its decomposition.

The carbon-nitrogen ratio is widened according to the data in the table, when nitrogen is added to the soil. Its lignin content is also increased. The addition of mineral fertilizers lowers the lignin content, or indicates the capacity of the bacteria to handle this sulfuric acid-resistant, organic residue to a higher degree when given mineral supplements, as in plot 2. Mineral manures thus reduce the level of organic matter in the soil, but do not deny a high rate of decay, or high degree of effectiveness of the smaller supply remaining.

The rate at which organic matter decomposes—taken as an index of its effectiveness—is dependent on the phosphorus and calcium, or fertilizers. Soils

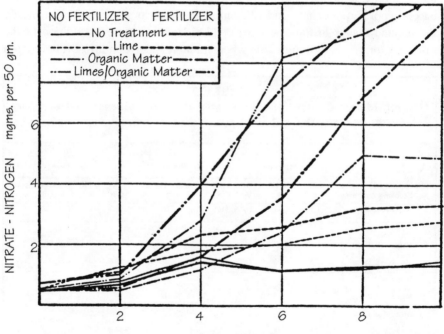

Figure 2. Nitrate accumulations or organic matter breakdown as influenced by long periods of fertilizer and no fertilizer additions when given different treatments in the laboratory. (Soils from Sanborn Field.)

from Sanborn Field taken into the laboratory for treatments and nitrate accumulation studies were selected from plots with and without fertilizers, and from those with and without limestone as comparison cases. In figure 2 are graphs which show the influence of the fertilizer treatments in the field on organic matter conversion into nitrates in the laboratory. It is interesting to note the effect of limestone in the laboratory when soils already had fertilizers in the field, but more significant is the difference where organic matter is added. The soil with fertilizer (mainly phosphates) makes the organic matter a better bacterial ration if nitrate production is the index. Where lime and organic matter are added together the soils act similarly in nitrate accumulation. With the blotting out of the difference between these two soils where the lime and organic matter are used alone, there is the suggestion that the difference in the latter was the equivalent of a lime effect. With the fertilizers carrying much gypsum such effect through the common factor, the calcium, seems logical reasoning.

The importance of the calcium is more clearly shown where the limed and unlimed soils were handled similarly. The much wider differences here are interesting, particularly the increased nitrates by lime additions to soil unlimed in the field, and that from the addition of only organic matter to the limed soil. These differences show clearly the significance of the calcium and phosphorus in the effectiveness of the organic matter in decomposition to liberate its organic nitrogen in the nitrate form usable by plants.

THERE IS NO SUBSTITUTE FOR SOIL FERTILITY

Crop growth depends on the usable stock of plant nutrients in the soil, or on that made available during the growing season. Crop rotations represent good soil management because they lessen the rate of plant nutrient depletion from the soil. Legume crops and grass sod covers are considered particularly effective in this respect because of the slower rate of soil nitrogen exhaustion in the former and because of the partial return of the nutrients to the field as animal droppings in the latter. Considered then purely from the crop-combination viewpoint, one can readily reason that any approach toward constant legumes or constant grass cover will approach satisfaction in terms of lessened soil depletion.

The short rotations of the cereals with newer legumes and the lessened soil tillage that have become popular in Missouri are built, in part, on this reasoning. A barley-lespedeza combination is 50% legume, for example. Their satisfaction, however, is not in the crop combination without soil treatment. It must take into consideration the facts, (a) that a soil is being exhausted by any crop order or arrangement under soil management that disregards fertility return; (b) that legumes are ravenous consumers of calcium, magnesium, phosphorus, potassium, and others; (c) that they are nitrogen fixers rather than only soil nitrogen consumers when generously supplied with calcium and the other essential soil bases; and, (d) that grass sod or non-leguminous crops, as well as legumes, are becoming less effective for their own preservation, or for forage feed production under increasing soil acidity, or, rather, base deficiency. In our present desire for more sod cover and more legumes, especially those growing on less fertile soils where we are expecting them to demonstrate their soil-saving results, we may well recognize many soils too low in calcium, phosphorus, and other nutrients before these crops can get even a "toe hold" in rejuvenating effects on the land.

The short rotation combinations of the cereals, barley or wheat with the more popular legumes, such as soybeans, Korean lespedeza, or possibly sweet clover, have been recently adopted for their soil cover value and their pasture extension over the main portion of the calendar. Their provocation of discussion of the subject of rotations has brought with it a recognition of the need for fitting crops to the fertility of the soil as well as into a particular rotation, or crop sequence. Soil fertility depletion has brought us to wheat and barley, in the place of oats, as nurse crops for legumes. This is true because the former two draw much of their needed nutrient supply in their young growth in the fall of the preceding year and lessen competition with the young legume during its first and the cereal harvest year. The oats are the poorer nurse crop, because they "get there first," so far as fertility is concerned, in the same season as the legume. They stay on well into the summer and make it difficult for legumes to succeed in the soil already sapped before they get underway at their delayed and possibly dry season. This shift to nurse crops that get off the land early is acceptance of the evidence of crop competition for a fertility that is getting to low to support both crops at the same time. It commonly places the blame on factors other than the soil, such as the seed or the season. The same declining fertility that has been too low for two crops at the same time has not only been obliterating the legume completely, but has also been lessening the productivity of the cereal, or non-legume, crop itself.

Fertilizers have been recognized as a serious need for the cereals. Superphosphate on wheat has been a safe investment in most parts of Missouri and has extended its effectiveness by establishing the legume nursed by the wheat. Now that drilling fine limestone with the wheat for clover the next spring shows liming beneficial on the wheat in the fall, as well as on the clover the succeeding year, we may remind ourselves that superphosphates of the past contained so much calcium that this element was perhaps doing much of the benefit for which phosphorus was getting the entire credit.

The graminiae, or grass family, on many of our soils respond to the calcium in limestone more than anticipated. Oats reflect the effects of lime additions in their growth, and even corn demonstrates the influence. Limestone serves to mobilize the nutrients into the plant. Fertilizers alone on corn mobilized phosphorus into the crop only one-third as effectively as when used in conjunction with limestone, according to chemical studies of the crop.

Parallel with the increase of phosphorus went an increase of calcium in the fodder crop. Barley, used extensively in conjunction with lespedeza, also reflects the beneficial influence of the phosphate and limestone treatments. Its earliness makes it the best of all nurse crops to draw its fertility the preceding season and leave a liberal supply for the lespedeza after the barley harvest. These cereal crops reflect the effects, particularly in composition of their forage, of the lime and phosphates to remind us that, in arranging rotations, the fertility of the soil may be a factor to determine the crop efficiency as well as the crop choice. They tell us that we cannot escape the fertility shortage by offering substitute crops.

Fertilizers for the legume in the rotation tell the same story as for the cereals, only with more emphasis. In our search for legumes which manage to grow on thinner soils, those so found must do so at a lower rate of fertility consumption by giving correspondingly lower yield rates, or smaller offering in the lower concentration of feed nutrients in the forage. Of course, lowered levels in all these are still better than complete crop failure.

Less calcium concentration in the legumes because of its depletion from the soil means less phosphorus mobilized into the plant. It means lowered concentration of protein in the forage, the tonnage yield of which means less protein yield per acre. So when these legumes make tons, it is well to appreciate the need for some test by which we can learn whether they are rich enough in such ash constituents as lime, phosphorus, potassium and others, or whether they are giving protein and items that do more than merely distend the consuming animal's paunch.

Declining soil fertility is reflected in natural vegetation by lowered yields within the same plant kind, or its disappearance and the incidence of such other kinds as can grow by using less nutrients from the soil. Such as make their advent into—and maintain themselves in—the ecological array at the lowest soil fertility level may be expected to be of low concentration in protein and ash. They must then be of low feed value as suggested by their refusal as forage by wild animals. The declining calcium, in particular, in conjunction with a moderately constant potassium level suggests the shift from the proteinaceous to woody plant composition even within the legumes. The same is indicated for the non-legumes to the point where wood production by trees may represent the extreme in calcium reduction in the calcium-potassium ratio. Naturally such also represents the lowest feed value as forage.

In disregarding soil fertility decline, represented by the large calcium depletion to give a narrowed calcium-potassium ratio, we are in danger of accepting forages of lower protein and lower mineral content irrespective of the crop combination into rotation schemes. Crops in short rotations, like any others, respond to soil treatments and are no escape from the facts that soil treatments can be beneficial to them.

"THE NATURAL" VS. "THE ARTIFICIAL"

It is all too easy to criticize a new landscape and to say that it is good or bad, beautiful or monotonous. This vagueness lies in the lack of any scale which would help us make evaluations . . . from a functional and aesthetic point of view. Good or bad must be related to something.

"Observing the changes of the rural landscape in most countries, the conclusion may be drawn that modern man is not only the conqueror but also the originator of *vacua*. The most representative phenomenon of contemporary landscape development is not the reclamation of deserts and sea lands, but the creation of vacant landscapes where there formerly existed a wholesome human environment.

"During the past 100 to 150 years our expanding technological civilization[1] treated much of the old land as if it was a vacuum. By systematic burning and clearing of vegetation, by levelling and rectangular subdivision, the *tabula rasa*[2] for man's exhaustive economy was prepared. By disregarding the ecological chain of biological, topographical, climatic, hydrological and social conditions of life in the landscape, land management led finally to the creation of what the Dutch call so significantly *the steppe of culture*. This widespread and monotonous type of landscape is characterized not only by being climatically and aesthetically unattractive but also by functional defects such as water and wind erosion, an unbalanced water cycle and declining soil fertility. The landscape constitutes today, a vacuum of a much greater extent than the potentially reclaimable water-covered and desert areas.

"The deterioration of the old rural landscape and the formation of an unsatisfactory environment on the new land cannot be explained just by what is called our *alienation from nature*, or the artificiality of the environment created by us.

"Let me state my interpretation of the notion *natural*. I believe that the notion of the *natural* condition of man does not express a past condition, left behind with the process of civilization, but a balanced relationship between human capabilities and natural forces which can be achieved, renewed or re-established at any level of civilization under the condition of the existence of wisdom, good will and energy.

"The modified landscape is not identical with the unnatural landscape. I believe the cultural landscape is by definition always a man-modified environment. And yet, in the past, the artificial increase of its functional values for man did not reduce its biological and aesthetic values . . . In the case of *the steppe of culture* the land use aim is narrowly conceived as a rule, and of temporary nature. It is limited to the maximum exploitation of soil fertility within the minimum of time, and this implies a readiness to abandon the land after the exhaustion of its reserves as if it were a mine.

"The fact that, in the past, communities realized that the land had to serve for generations as a cultivable and renewable source of life, as a permanent place of habitation, work, celebration, movement and rest, is the main cause for the sensation of environmental wholeness and aesthetic satisfaction which we experience in a preserved, traditional landscape. As planners for the future of the land, however, our cardinal aim should not be to preserve, or to emulate such landscapes, but to regain the scope which originated in them.

"The most obvious aim for the development of new land lies today in the efficient and intensive production of food and raw materials for the needs of populations. This is the quantitative aim and it should be brought into harmony with the broader spatial and temporal aims of development which we generally characterize as qualitative. It very often seems, though, that the quantitative aims emerge from the immediate short range considerations, and that the qualitative aims are related in the future."

In his view of the future, Dr. Glikson cites: "The first principle: the cultural landscape should be planned for an optimal sustained level of soil fertility. Consideration of the needs of intensive production on one hand, and of soil protection, vitalization of microbial soil life and preservation of the water cycle, on the other hand, have all to be taken into account and to be balanced with each other. Once this is accepted (we) make use of ecology as an applied science. A balanced relationship between man and the land means the achievement of a new ecological climate in the landscape, which is arrived at by a conscious and rational effort of man . . . Sustained soil productivity as a cardinal purpose of cultivation should become a major landscape-shaping factor overriding consideration of a more temporary character.

"The second principle: the standard of living of a modern rural population in industrialized countries must be adjusted to that of the urban population, though rural life will always represent a different kind of life.

"The third point: the general increase of leisure and of demands for more first-hand experience of different environments, combined with the availability of faster and cheaper means of communication for the masses, have revolutionized the mobility of the whole population of industrialized regions. Today this free, so to speak, uneconomic movement is still in its very beginnings, but is rapidly gathering momentum.

"If there is any room at all for an optimistic view toward the future, one may forecast an uprise of the urge for direct biological and social contracts resulting, among other things, in a return movement of urbanites to the landscape. We should not regard this development as a new and overwhelming danger to the landscape. . . . It would involve the rediscovery, reconstruction, recreation of the cultural landscape with the active participation of the urban population. We should meet this development by planning an environment both accessible and hospitable to what may be called the recreational movement, recreational in both passive and active sense of the word."

SALT DAMAGE TO SEEDLINGS

Even though we know that saline and alkaline soils damage the germinating seeds in arid regions, we seem to be little concerned about a similar injury from salt fertilizers drilled directly with the seedlings on soils of humid regions.

Yet such damage was demonstrated on Sanborn Field at the Missouri Experiment Station nearly 50 years ago. Apparently that unpleasant fact remains hidden in the field man's notebook, written carefully in longhand but never printed in publications emphasizing large yields from particular soil treatments or plant varieties.

When Professor J.W. Sanborn laid out the plans for Sanborn Field in 1888, he wanted to learn what will result (a) when the soil is cropped continuously to wheat; (b) when the entire crop of straw and grain is removed; and (c) when the fertility equivalent of only the nitrogen, the phosphorus, and the potassium in the entire crop of 40 bushels per acre is put back in the form of commercial fertilizers. For the salts of that equivalent, he started with sodium nitrate (Chile saltpeter), acid phosphate (16%), and muriate of potash drilled with the fall seeding.

The early successful wheat yields were heartening, No one had anticipated a later demonstration of salt damage to the extent of complete destruction of the stand. That had not been put into the plan. It was merely an observation recorded in the workman's notebook and not in the "Official Records," That salt damage should occur was scarcely expected. There had been no previous injury from the fertilizer's contact with the seed, even at that heavy treatment, since the outset nearly 15 years before.

But after an unaccounted, yet significant, share of the virgin organic matter of the soil had fed the highly stimulated microbial fires to burn that supply to a lower level of shock-absorbing capacity for salts, there occurred destruction of the stand following the seeding. There was at that time no thought of placing the fertilizer away from the seed when the cause of the trouble was suspected. Instead, as a test case, the application was split into two parts; one-half the amount drilled with the seedling, the other half applied as a surface dressing the following spring.

Later on in the history of that classic plot, similar trouble was experienced again. A more extensive number of fertility elements, such as the salts of the trace elements, were included in the treatment. This called for the practice of fertilizer placement to avoid the salt damage on the soil, which was now very low in its content of organic matter.

That early problem of damage by salt fertilizers brought about glasshouse tests of many oncoming kinds of fertilizers to learn their "maximum permissible concentration" for different crops and different soils. (The MPC used was for supposedly acceptable poisons in food.) There have accumulated many notebooks reporting data by advanced students doing "special problems" and graduate theses. Among them are many interesting reports. These emphasized the damages to the legume crops, even before damage to the wheat, which was usually their "nurse crop." Legumes were recommended for their "natural" nitrogen fixation as more desirable than nitrogen salts fertilizer.

One single table [in a notebook] from those student records illustrated the wide range between danger and safety exhibited in the fertilizing materials and in two kinds of legumes tested in one trial.

The problem of damage by salt fertilizers started glass-house tests later, and many notebooks containing data by advanced students have accumulated over the years. Among them are many interesting accounts of damage to legume crops. Legumes were emphasized for their "natural" nitrogen salts. The comparison illustrated the wide range between danger and safety exhibited in the fertilizing materials and in two kinds of legumes tested in one trial reported by the data in the comparison. In those years when lime and phosphates for legumes were first emphasized for agricultural uplift, and the service by lime as plant nutrition rather than improved soil environment was a theory, a glasshouse test was run on the possible damage to red and sweet clovers as reported for tested materials.

It was significant to note from the differing percentages for emergence of the sweet clover that this "hard-seed" crop was erratic in its germination and seedling survival. It was not so regular in responses as the red clover. Yet the seed was not "hard" enough to escape damages to the stand (as was true for red clover) when (a) the alkaline hydrated lime tailings, (b) the basic nitrogen salt of ammonium phosphate, and (c) the slowly-transforming calcium cyanamid were put in contact with it. The data is certainly not erratic in reporting those salt damages from soluble fertilizers.

The more regular data by red clover showed that superphosphate (not inspected for water solubility) was not seriously disturbing to emergence. In fact, this seemed to be less so in the data for 250 pounds/acre than was the superfine limestone for 500 pounds/acre. When that phosphate was combined with potassium (a water-soluble salt) the emergence of the red clover was cut to reduce the stand of this legume plant seriously. But when phosphate was combined with nitrogen in dibasic ammonium phosphate, the damage by seed contact was disastrous.

While calcium in the limestone (as a carbonate rock) and in the superphosphate (as soluble gypsum) was apparently without damage to the seedlings of either legume, its combination with nitrogen, as occurs in calcium cyanamid, was the one test with most disastrous effects on both the red and sweet clover.

That cyanamid was most damaging was to be expected when this slowly transforming compound was put in contact with the clover seed. It undergoes chemical changes when moistened. Time is required for about four steps in those changes by which the nitrogen separates completely from the calcium. However, alkalinity by the latter encourages formation of the dicyandiamide compound, which is damaging to the seedling. By applying cyanamid ahead of seeding, the danger can be avoided.

Though sad experience has told us long ago about possible crop damage from salt fertilizers in contact with seeds, that fact remained hidden from many. It took years to avoid the trouble through so-called "fertilizer placement" by special machinery, and by building up the soil fertility in advance of the seeding, especially by using the fertilizer ahead on green-manure crops, or via composts and other kinds of fertilizing organic matter. We learn by post-mortems more often and more sadly than by prophecies and reported facts.

GARDEN SOILS AND BIO-GEOCHEMISTRY

Though most of us take our soils for granted, there is an increasing interest in the effects by the soil and its original rocks (a) as the rocks are weathering to develop the soil, and (b) as those materials and processes may determine the healthful, or possibly poisonous, quality of the food grown thereby.

Bio-geochemistry is a relatively young science. It is the study of the ecological pattern, that is, the distribution of life forms on the earth in relation to the chemical potential of the rocks and minerals for soils to grow their food support. This science gave an early impetus to the study of plant compositions for their help as indicators of concentrated ore bodies for mining. Because rocks make soils, and because the fertility of those as nutrition, for food crops, determines the biotic geography, the geologists are concerning themselves with the nutritional qualities of food supplies according to their contents coming from soil minerals. They emphasize the trace elements in particular.

Professor H.V. Warren of the Department of Geology of the University of British Columbia has given us the chemical analyses of the soils and of the

Table 1
Trace Elements in Garden Soils and Plants (Amounts as Parts Per Million, Ash of Plants in Parentheses)

"A" GARDEN

80 mesh fraction of soils	No. of samples	Copper	Zinc	Molybdenum	Lead
Aqua regia extraction	5	11	71	3	84
Acetic acid extraction	5	4	21	N.A.*	11
Whole vegetable sampled	9	7(55)	54(391)	6(44)	7(44)
Edible portion only	5	6(60)	46(421)	3(31)	5(37)

"B" GARDEN

Aqua regia extraction	5	20	58	15	5
Acetic acid extraction	5	0.2	16	N.A.*	4
Whole vegetable sampled	7	3(25)	37(257)	9(73)	6(43)
Edible portion only	3	2(27)	19(250)	10(90)	2(33)

*N.A. = not analyzed.

vegetables (both total and edible parts) for the four trace elements, copper (Cu), zinc (Zn), molybdenum (Mo), and lead (Pb) of two gardens. They are within 200 yards of each other. Both are soils developed from the same rock formation. They are in a community of multiple sclerosis. "Lead has been suspected as possible cause, but the case against this base metal has never been proved."[1]

His data is given in table 1, with quantities of the elements reported as parts per million in the dry matter and in the plant ash. Only that part of the soil which passed through the 80 mesh screen was extracted by the two acids, one extremely strong, aqua regia, and the other very weak, acetic acid.

Since it has been established that "oven-dried (moisture-free) food products normally contain 0.1 to 1.0 ppm of lead," and "in the ash these values range from 2 to 20 ppm"[2] it is evident that "anybody who eats food from these gardens would obtain from five to 15 times his normal intake of lead.

"There is much to be learned yet before we have 'normals' for trace elements in foods; also before we can risk any statement for 'normals' of them in soils. We are not certain as to what part of a trace element in the vegetable is taken into the system, and what passes through the digestive tract unchanged. This is illustrated by calcium in an oxalate compound in some vegetable greens, like those of the goosefoot family, while in the mustard family this indigestible compound is not so prevalent."[3]

"Until the chemical forms in which a trace element is present in any food can be determined, and likewise the digestive peculiarities of the eater of any food, it is not possible to assess the nutritional significance of abnormal amounts of such elements as lead, selenium, and molybdenum in vegetables.

"There is abundant evidence," Professor Warren reports, "to show that an anomalously high or low trace element content in a rock formation is reflected in comparable anomalies in the soils derived from those rocks. Soils, in turn, pass on their trace element variations to the vegetal matter growing on them, even when these trace elements are not thought to serve any known purpose in the plant physiology.

"Vegetal matter used as food may not contain as high concentrations of some trace elements as some vegetal matter not normally thought of as food. Nevertheless, most foods can be shown to reflect the trace element peculiarities of the soil and of the rocks to which they are related. An overabundance of trace elements, such as mercury, lead, cadmium, or selenium, whether eaten in vegetal or animal foods, can seriously affect the health of humans. Deficiencies of the elements, such as copper, iron, manganese, zinc, and molybdenum, may also cause nutritional problems. "It would appear, then, that any diet marked by imbalance of one or more trace elements, or, indeed, in any macro-nutrient, must be viewed as a potential cause of ill health. This would seem to be true whether applied to primary foods, such as oatmeal or cornpone, or to secondary foods, such as milk or cheese."

Much is yet to be learned by those professionally concerned with nutrition and health if they will consider the possible contributions by the many other sciences not so directly related. Shall we not agree that all of us, whether scientists or not, are individually concerned and responsible for our health, dependent as it is in each case on nutrition in terms of what we take and what we avoid? If we are to survive by means of the natural chemo-dynamics that operate the assembly line of creation from rocks to soils to foods and to life, then all sciences may well become more concerned about what they can contribute toward making us more fit for it under rapid population increase.

LESS SOIL ORGANIC MATTER SPELLS LOWER FORM OF VEGETATION

In 1945 the microbe *Streptomyces aureofaciens*, which produces the antibiotic aureomycin, was isolated by the late Benjamin M. Duggar, Ph.D., from the soil of experimental plot No. 23 (which was in continuous timothy) of Sanborn Field at the Missouri Agriculture Experiment Station.

Noteworthy is the correlation between the climatic setting, soil conditions, the associated or competitive crops on that untreated plot and the unique power of self-protection against bacteria exhibited by the timothy. This is particularly important, if we dare classify *S. aureofaciens* as a climax crop of that microbial group, due to its greater presence in the particular plot, in the particular soil type, and in a particular degree of soil development under the climatic forces, all reflected as unique among the many soil samples collected from the 50 states and several foreign countries. We need to remind ourselves that any climax crop, whether microbe, plant, animal, or man, surviving more fitly than other similar crops, has a certain group of other life forms beneficially associated with it.

The American bison was a climax animal crop of the plains and the prairies. It grazed on certain plant species, notably the many grasses and legumes. Certain microbial life forms were prominent in the soils of its habitat. These included Azotobacter (and Clostridium) which could use gaseous nitrogen from the air to build the protein of these free-living cells. There was also the soil microbe Rhizobium, building its protein similarly but by symbiotic relations with legume plants and their root nodules. These soil microbes were starting with atmospheric nitrogen, getting energy from decaying organic matter, and building the protein foundation for the brawny bison of pioneer days on those grassy areas and for the herds of cattle and sheep that took over that climatic setting later.

Since every form of life has survived naturally where it has been most fit, it has been a challenge to continue observations, isolations, and tests of this antibiotic producer in Missouri soils in order to catalog its behaviors in relation to climatic range, soil types, and managements of soils and crops supporting it as a climax soil microbe.

The first significant observation was the fact that the initial soil sample giving the choice specimen of *S. aureofaciens* (No. A 377) as producer of aureomycin, came from a plot with no soil treatment given to continuous timothy with its annual crop removal. In contrast, this microorganism could not be isolated from the plot alongside (No. 22), also given to continuous timothy with its annual crop removal, but with the annual application of the equivalent of six tons of barnyard manure per acre.

Yet "the only soil we have ever seen which was really 'loaded' with *S. aureofaciens* was Duggar's sample 65 which indicates it was collected from below the bluegrass and on the roadway adjacent . . . (Sanborn Field) . . . , unplowed about 50 years. This sample yielded a total of 18 *S. aureofaciens* strains. Besides the original A 377 which came from plot 23 in Sanborn Field, two isolates were obtained, one from plot No. 10 and one from plot No. 9," according to Edward J. Backus, Ph.D., now head of the Microbiology Department, Biochemical Research Section, Lederle Laboratories.

The originally fruitful sample from plot No. 23, giving strain A 377, was taken when a dense crop of the "weed" *Andropogon virginicus* had taken over seriously enough to grow a ton and a half of its vegetation after hay-cutting in

June.

As a test of the possible association of *S. aureofaciens* with this plant, commonly called "broom sedge," samples were collected in the spring of 1962, again from plots 22 and 23, Sanborn Field (plowed and reseeded in autumn, 1961). Also, samples were taken from a pasture on the farm of A. Sapp, Ashland, Missouri, representing Putnam silt loam, similar to Sanborn Field—one where broom sedge was very prominent and one where it was absent and drainage was somewhat impeded to give a mossy growth. The pasture has been in unplowed sod for more than 20 years.

Again, the two samples from Sanborn Field in 1962 were reported for duplicating their original condition, namely, two strains of strongly pigmented *S. aureofaciens* isolated from the untreated plot No. 23, but none from the adjoining manured plot No. 22, both in timothy since 1888. From the samples of the same soil type, that is, Putnam silt loam, of the older unplowed pasture at Ashland, Missouri, there was isolated from the area "growing *Andropogon virginicus* and not plowed since 1920 one strain *S. aureofaciens,* this being culturally almost identical to Dr. Duggar's original A 377 strain." From the Ashland soil "growing no *Andropogon virginicus*, one strain of *S. aureofaciens* was found, being a more intensely pigmented type than A 377," according to the report by Dr. Backus of the Lederle Laboratory, through whom the results of the test were obtained.

Isolations were made from other samplings of untilled Putnam silt loam as Dr. Backus reported. "We did find single strains of the organism in each of the three samples which we collected (October, 1948) near Centralia, Missouri. One of these samples had an Andropogon cover in a railroad prairie remnant, while the other two were associated with bluegrass sods."

Since Sanborn Field seems to support the antibiotic-producing *S. aureofaciens* under conditions of low soil fertility, constantly threatening invasions of "weed" crops which animals refuse to eat, as undesirable invaders demanding tillage for their exclusion, and under old pastures and roadways with little or no crop removal, it would appear that this antibiotic-producing soil microbe is an indicator of declining soil fertility for farm crops.

Since this bit of soil life is of the fungus order getting its energy from resistant organic matter, and then producing its antibiotic for self-protection against so many other microbial forms injurious to animals and man, shall we not consider this laboratory-classified, lowly form of soil-supported life an indicator of the soil's need for uplift of inorganic fertility and supply of organic matter in order to bring its crop output to higher nutritional values? In our past agronomic experiences, there are hidden reports telling us that when our soils have less organic matter reserves they can produce only vegetation of lower nutritional value for higher life forms, and may even be producers of powerful poisons.

4

THE ANATOMY OF SOIL FERTILITY

FERTILE SOILS FOR SELF-PROTECTION BY CROPS

When fungus and insect troubles attack our crops we think first of counter attack with copper and arsenic and other poisonous weapons. Instead of viewing such troubles as an immediate call to fight the invader, should we not think of them also as a call to make future crops healthier and thereby more resistant?

Why not view it as a case of partially starved plants that are too weak to ward off attacks by these microforms of plant life, and one in which the insects are merely serving to hasten the disposal of what is already near death? Better knowledge of the physiologies of plants and insects is bringing us to think more about treating the soil to help the plant set up self-protection by growing itself more healthily. At the same time the more healthy crops are very probably growing into bigger yields and better quality in terms of nutrition.

For protection of ourselves against many diseases, the inoculations injected into our bodies are usually carried in the form of proteins, such as blood serums, probably of animal origin. Has it ever occurred to you that plants are protecting themselves by means of a particular kind of protein, or a higher concentration within them of elements of fertility, especially nitrogen, which is almost synonymous with protein?

That plants can ward off insect attacks because they are healthier when fed with more nitrogen was recently demonstrated. Spinach was put under study for variation in its vitamin content as this synthetic product by the plant might be related to varying levels of nitrogen in the soil. Each of these levels was combined with a series of levels of calcium. Both calcium and nitrogen are connected with protein syntheses by plants. This vegetable crop demonstrated a pronounced susceptibility to the attack by the thrips insects (*Heliothrips haemorrhoidalis*) when it was grown on soils low in nitrogen. But it was immune to injury by these insects when it was growing on soils to which nitrogen had been applied more generously.

Within the lower levels of nitrogen that were inviting the attack by thrips, the plants were injured less as there was more lime, or calcium, combined with the particular application of the nitrogen. Here were clear-cut demonstrations that the well-nourished, and therefore healthy spinach plants, were warding off attacks by this insect (that eats the green parts of the leaves) leaving them with clear areas. It was only those plants which were partly starved for nitrogen (in spite of the help by calcium) that fell as victims of this pest.

A fungus disease of soybeans demonstrated a similar failure to attack plants when they were growing on soil that provided them with more calcium. An increasing percentage of the young soybean plants was attacked in early growth to the degree that calcium was available in the soil supporting them. As more clay was present in the sandy soil to give the roots more contact with this source of exchangeable calcium, less plants were taken by what suggested itself as a trouble similar to that commonly called "damping off." This beneficial effect from more clay in the soil was evident though the clay has very low calcium saturation, as revealed by its pH figure of 4.4. Good protection from more calcium and injury from less calcium occurred when the plants were all grown in such close proximity that the same invading shower of fungus spores should have reached all of them. And yet, within the series of increased clay, and therefore increased calcium, there was represented an infection of almost 100% for the calcium-starved plants, and immunity of equally high percentage (or what amounted to perfectly healthy plants) for those with the largest amounts of calcium available in the soil.

When nitrogen is deficient in the soil system, when calcium stands first on the list of nutrients which humid soils must be given for production of protein-rich, nitrogen-fixing, leguminous plants) and when lime exercises pronounced synergistic effects in mobilizing nitrogen and other nutrients from the soil into the crops—are we not justified in believing that nitrogen and calcium (so closely associated with the production of protein by plants) exercise their protective effects for plants by way of the protein?

Observations on fertilized wheat having less Hessian fly trouble than wheat not fertilized were reported by the Doane Agricultural Service of St. Louis, Missouri for adjoining fields in Illinois several years ago. One field in a four year crop rotation had been limed and fertilized. It was observed to have very little damage from these more common wheat insects. It produced a yield per acre of 30 bushels of this cereal grain. The adjoining field was farmed to wheat by the same men. It had, however, never been limed, nor rotated in its crops. It had regularly been put to wheat with no fertilizer or other soil treatments. On this field the Hessian fly brought severe damage and the yield of grain was only six bushels per acre.

Similar observations were made in another area of that state. Wheat seeded after oats on land given no soil treatments had a heavy infestation by Hessian fly, and the crop yield was only 12.3 bushels per acre. On a similar kind of soil, but one well fertilized on the same farm and under the attention of the same farmer, there was very little damage by the Hessian fly. There was a yield of 30 bushels of wheat pre acre.

These contrasts of carefully measured yields, and similar general observations reported by other folks, have indicated that soils built up in nitrogen—and fertility in general—mean not only more grain yield through healthier plants, but also self-induced immunity. It was this immunity that contributed also to a bigger harvest.

Just how the plant protects itself against fungus disease and insect attack has not been placed under the category of any single factor, as has been the case for the human body, where much of the protection resides in the blood serum or blood protein. But when it is the actively growing plant that is healthy, when the stream of life does not flow from one fat globule to another, or from one carbohydrate molecule to another, and when cell multiplication and life itself go through the protein for which nitrogen is the distinguishing constituent—isn't it time to think about more organic matter for more nitrogen in the soil accompanied by more mineral fertility as protection for our crops?

Isn't it time to think of this immunity factor as an aid to our growing a larger volume of production with higher nutritional value? If insects attack our food crops, might we not raise the question whether the plants are truly healthy for their own good or even for the good of us who eat them? Would plants have survived in the course of evolution had they—in the early, negative stage of their life—always been eliminated by insect attacks?

Perhaps insects and diseases are merely militating against our attempts to grow on fertility fringe soils, plants close to starvation because of deficient soil fertility. If more, or possibly special, protein grown by the plant within itself means more self-protection by the crop against disease and insects, or against what in our frustration is so often called "bad luck," perhaps we can get around to believing that by having more nitrogen, more calcium, and more fertility in our soils to support the plant's struggle to feed itself, we can bring "good luck" to ourselves in the form of healthier, bigger and better crops.

SOIL FERTILITY A WEAPON AGAINST WEEDS

It is becoming a more common complaint that "Broom sedge and tickle grass are taking our pastures." Others say, "Those weeds have moved in from the South." All this has happened in spite of statutes against the spread of noxious weed seeds.

Broom sedge has come into prominence, especially as a leftover for winter notice, because the cows refuse to eat it. Tickle grass comes under the same category. That refusal has been the cow's way of telling us that such plant growth does not deliver enough feed value for her trouble of eating it.

Sanborn Field, at the Missouri Experiment Station, is giving its wise answer to the question of how to get rid of these noxious plants. It tells us clearly that broom sedge and tickle grass have come in because the higher level of soil fertility required for more nutritious feed plants has gone out. Rebuilding the soil is the answer.

Pasture farming with continuous grass must already have been in the mind of Professor J.W. Sanborn when, in planning this now historic field, he included two plots of continuous timothy. One has been given six tons of barnyard manure annually, the other no treatment, ever since the field was started. The manured plot has always been a fine timothy sward with early spring and late fall growths, and a quality hay crop of two and one-half tons every summer. The one with no treatment has been getting so foul with growths other than timothy as to require plowing and reseeding about every five or six years. This called for the same treatment, for uniformity's sake, of the companion plot, even though weed-free. broom sedge soon takes over the no-treatment plot as the major troublemaker.

These plots tell us that the trouble is not in the particular "invading power" of the broom sedge, even if its fluffy, wind-born seeds travel profusely and go everywhere. Both plots are equally invaded by the seed of this pest. However, broom sedge appears on only the untreated plot, cropped to poor hay now much of the time. It grows right up to the soil line dividing these two plots. It does not cross that border. None has grown on the manured plot, always covered with a dense growth of good timothy. Other plots equally as low in soil fertility on this field are gradually being taken over by broom sedge. Those with higher levels of productivity are free of it.

Professor Sanborn says: "Fertilize the soil so it will grow grasses that make nutritious feed and these troublesome plants of no feed value stay out." "Poverty bluestem," as broom sedge is sometimes called, is a mark of the kind of farming that fails to consider the fertility of the soil as the foundation of it. This pest has been coming more and more into what has been "pasture farming" because past cropping to grass, like to any other crop, without soil treatments, has brought the fertility down to where crops of feed value are failing and only those woody growths like this poor bluestem the cows reject are all that can be produced.

Tickle grass is also invading Sanborn Field. It has been taking over the last half of the six year rotation of corn, oats, wheat, clover, timothy and timothy where no soil treatments are used. The plots alongside given manure, phosphate, and limestone are free of it right to the straight line separating the plots and treatments.

This poverty grass means poverty of soil fertility and thereby poverty in crop output of yields and of nutritional values. No other crop can grow luxuriantly enough on such low fertility to drive this pest out.

This weed has come in when the rotation was as long as six years. This is one claimed by some to be helpful for better yields and crops. It certainly has not been such. It has simply rotated the fertility of the soil out so much faster, that tickle grass now comes in as major cover during those three years when there ought to be clover and timothy. Crops requiring no plowing have been rotated out and tickle grass has been rotated in.

Chemical analyses of these pest crops confirm the suggestions by the cow and Sanborn Field. As pounds per ton, broom sedge contained 2.14 of calcium, 2.18 of phosphorus and 88 of crude protein. For tickle grass the corresponding figures were 3.04; 1.9; and 69. In marked contrast for a mixture of clover and grass from fertile soil and at similar growth stage the corresponding values per ton were 26, 4, and 181 pounds.

Nature's pattern of different crops growing on different places is the pattern of different levels of soil fertility nourishing them and correspondingly a pattern of different feed values in them. Declining soil fertility is pushing some crops out and letting others come in. According to these simple soil differences, then, we can drive out broom sedge and tickle grass by lime, manure and other soil treatments offering better nourishment for better crops that will keep these pest crops out. That broom sedge and tickle grass should go out when you put soil fertility in is merely the converse of the fact that they come in when the fertility goes out.

WEEDS, AS THE COWS CLASSIFY THEM

Weeds are commonly defined as an undesirable and worthless crop. But, when we think a bit deeper than the crops, and when we study the soils under them, we must use another definition. When weeds are left to grow in the pasture, that is, grow bigger because they are disregarded by the cattle, then we must define weeds as a crop so poor in nutritional values, because of the poor soil under it, that a cow has sense enough not to eat it. Have you ever thought that weeds left in the pasture are pointing to the need to treat that soil with fertility additions if it is to keep good feed crops growing? The cow has never learned the names of plant species nor memorized the "manual of weeds," but she knows the nutritional quality of the vegetation according to the fertility of the soil growing it. She demonstrates that very accurately whenever she has a chance to choose.

More than a hundred head of beef cattle gave such a demonstration on the Poirot Farms, near Golden City, Missouri. What those cows called "weeds," namely, a worthless crop as they judged it, was in decided contradiction to our customary classification of certain plant species as weeds. They refused and disregarded bluegrass, white clover, and even some soybeans in a virgin prairie that had never had any soil treatment. For them the bluegrass and white clover were weeds, namely, plants that they had sense enough not to eat, because the herd marched right across this large field and through the gate on the opposite side to eat what had grown up in the previous year's cornfield left unused because of labor shortage.

On that abandoned field of choice forages as demonstrated by these cattle, there were only plant species which the bulletins and books call "weeds." There cockle burs, nettles, plantain, cheat, wild carrots, butterprint, wild lettuce, berry vines, and a host of others. Strange as it seems, all were eaten by the cattle and kept down to a short growth during the season while the adjoining virgin grass area, which they traversed daily for water, grew taller and taller.

"That temporarily abandoned but well fertilized cornfield in weeds," you would be compelled to say, "was good pasture in the cows' choice despite plant species that we call worthless but the cows select, because of the higher fertility of the soil growing them." For the cows, the adjoining heavy sward of bluegrass and white clover was a worthless crop and thereby they classified it as weeds according to this demonstration of the herd that was expected to eat them. The cows defined a weed, then, as any plant, regardless of species or pedigree growing where the fertility of the soil is too low, or too unbalanced, to let the plant create what is nourishment for the beast.

Not so long ago, an able botanist surveyed and listed the plant species in the flora on the western plains where a large herd of cattle was grazing. He reported 65 different kinds of plants in the herbage and, strange as it may seem, not a one of them was refused or left untouched by the cattle. On those more fertile, less leached soils under lower rainfall, the cows literally made a clean cut of the forage, irrespective of plant species. Differences in the so-called "palatability" of the plants according to their names are seemingly unknown on those mineral-rich soils.

So when we characterize certain plant species by calling them "weeds" or by saying they either are or are not "palatable," we are giving more or less academic definitions. These definitions apparently need drastic revision if they are to agree with what the cow "calls" them in her classifications.

When we talk extensively about going from corn farming, with so much plowing and its erosion, to a grass agriculture for less soil loss, the cow's judgment of that shift may well be consulted. We need to make that decision with the cow sitting at the conference table as the foremost member of the planning committee. She will recommend that we forget the idea of merely shifting to another combination of crops that we expect her to eat. Instead, she will point that erosion has increased because of declining soil fertility, and that such lowered fertility will make any combination of crops merely weeds as she defines them. She will tell us that it is not any particular system of agriculture but the higher fertility of the soil under it that makes any kind of crops worthwhile feed for her. Going to a grass agriculture that will feed and not fool the cow calls for improvement of the soil fertility first and then the selection of the crop combinations second.

The cows on the Poirot Farms voted against a grass agriculture of bluegrass and white clover on soils of neglected fertility. On the contrary, they preferred well fertilized land under corn farming and the temporary neglect provoked by economic pressure, that could grow only the crop combinations which quickly seeded themselves and we would call weeds. Cows know their crops, even weeds, better than we do. They judge crops, not by pedigrees, species names, nor even tonnage yields, but by the fertility of the soil growing them and thereby the nutritional values as cow feed. The cows recommended more attention to fertilizing the soil and less to fighting the weeds. As we make the soils more fertile in our pastures, we shall have more nutritious forage and fewer weeds as the cows classify them.

WEEDS SUGGEST LOW NUTRITIONAL VALUES

Earlier, we discussed the widespread acceptance some 25 years ago of Korean lespedeza or bush clover as an "acid-tolerant legume." It was claimed to be less expensive than alfalfa and red clover on the grounds that "One does not need to lime the soil to start this new crop." Yet these early claims made no mention of the improvement that could be shown in the new legume crop—particularly in thicker stands to eliminate weeds—if it were given rock mineral treatments.

The earlier trials with the lespedeza brought marked results as the soil fertility was brought into higher levels by the addition of rock phosphate and limestone. Through this combined treatment the weeds were eliminated, giving a pure stand of bush clover. Most important, however was the increased yield

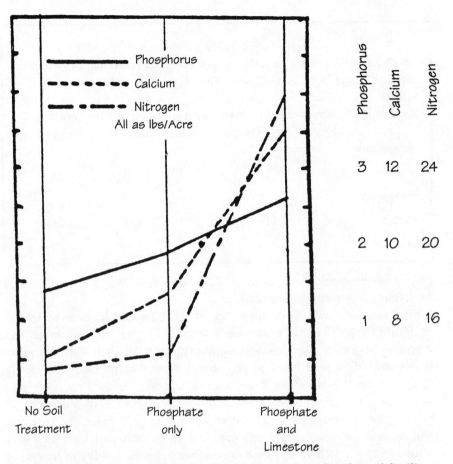

Figure 1. The uplift of nutritional values in Korean lespedeza by soil fertility treatments, which also eliminated weeds, would not have been obtained had the latter resulted from the use of herbicides.

of bulk hay and crude protein. The legume was a welcome supplement in animal rations predominantly carbohydrate.

The experiments were demonstrations of the natural law, showing that the incidence of weeds was due to man's failure to nourish a crop properly at its outset.

The experiments were also a demonstration of the fact that, had we chosen to eliminate the weeds by herbicides, the result would not have been an increase in nutritional values (figure 1) such as that obtained by phosphate and limestone treatments, which were simultaneously and incidentally the eliminators of the weed pests. Herbicides do not raise feed values of the crop, especially since plants take up poisonous organic compounds, even those of a greater degree of complexity in chemical structure. Improvement of the nutritional service of a

Table 1

Much Larger Increases in Yields of Lespedeza Hay Resulted From Better Balance by the Two Soil Treatments Than From One Only.

Soil Treatment	**Percentage Increases per Acre's Harvest (in Terms of—)**				
	Hay	Crude Protein	Calcium	Nitrogen	Phosphorus
Phosphorus only	5	17	23	6	16
Phosphorus and Limestone	83	113	85	114	75

crop is more valuable than waging a "fight" on weeds. Why not use the funds for fertility, rather than for poisons?

Most significant was the demonstration of the soil's need for more balance in its fertility supply. It has been widely recognized that for grain crops and legumes phosphorus is the foremost requisite on most Missouri soils. But many tests restricted by such belief to phosphorus alone demonstrated very clearly that, even for this reportedly "acid-tolerant legume," calcium (possibly also magnesium) as a plant nutrient applied by limestone serves to do much more to improve the harvest. This improvement was not only in vegetative bulk, but also in more protein calcium, nitrogen, and phosphorus mobilized as higher organo-inorganic feed values when the combined mineral fertility treatments of limestone and rock phosphate were offered.

It was the soil treatments, not the foreign pedigree, which made this imported legume such significant summer grazing in so many needed places. Phosphate alone gave increases, but the effects were small as compared with those obtained by the combination of the two treatments. Percentage increases for the effects by limestone are shown in table 1.

We are slow to appreciate the fact that the growth of a feed crop, like the legumes, requires a truly "living" soil with active microbial life in its many forms. We need their action of decaying organic matter to mobilize large quantities of the fertility elements that animals and man must have from the soil source. We, seemingly, still take the soil for granted.

Although we have gained in yields per acre in some aspects, such as high starch output per acre for fattening services, nevertheless, the declining health and reproductive aspects of both plants and animals call for better nutrition in proteins, and all that goes with them, via properly fertilized soils, rather than those dosed by powerful poisons in our war on weeds, fungus diseases, insects, and other pests commonly serving naturally to eliminate higher life forms not fit to survive in the competition.

Unfortunately our minds, like many good books, remain closed and nature's facts stay hidden from us all too long.

PROTEIN IS PROTECTION

It was an advertisement for tobacco which a few years ago kept reminding us that "Nature in the raw is seldom mild." That is a way of saying that not only the strong within a species survive at the expense of the weak, but also that the weaker among one species may fall victims to other species.

That is true not only when we know that weakly nourished plants fall victims to bacterial and fungus attacks, but it is true for poorly nourished plants falling victims of the attack by insects. Perhaps you have never thought that our increasing troubles with insects may be the result of the declining fertility of our soils. Such are the facts demonstrated by recent experimental work.

Spinach was grown at the planting rate of two per pot on soils in which the clay had been given uniformly specific amounts of all the nutrient elements except calcium and nitrogen. These two elements were arranged in a series of 5, 10, 20 and 40 milligram equivalents (ME) per plant of nitrogen and each quantity of this was put in combination with the corresponding series of amounts of calcium.

During the mid-mature stage of growth of the plants, there was a scourge by the thrip's insects. These eat the green portion of the leaves to leave clear areas of the skeletonized tissue. The severe attacks on some plants, and the complete abstinence by the insects on some others, marked itself out in a pattern—according to the soil fertility applied in varying amounts of nitrogen and of calcium. This arrangement of the insect attack on the plants, according to soil fertility, is all the more striking since the treatments were replicated ten times, and the arrangements of the insect behaviors were thereby repeated ten times.

Insect attacks were numerous on the spinach plants given only 5 and 10 ME of nitrogen. There were no attacks by the insects on any of the plants given 20 to 40 ME of this nutrient which is valued for its services in protein synthesis and its ever presence in the chemical makeup of protein. When we remember that calcium, or lime, is a necessary treatment on humid soils to grow protein producing legumes, it is especially significant to note that even when the insects attacked those plants given the smaller amounts of nitrogen, those attacks were less severe according as those small amounts increasing the calcium as additional soil treatment.

Here is the suggestion that the nutrition of the plant, that is, the better balance in the fertility of the soil for production of more protein in the crop, represents better protection by the crop itself against insect attacks. "Strength" of the plant in terms of guarding itself against insects is also strength for reproduction when we remind ourselves that it is only protein that can reproduce itself. These plants using more nitrogen and more calcium to make themselves more proteinaceous were growing, or building, their own defense within themselves against the thrips insects. Perhaps you might say "They had built up a resistance" to this trouble. This resistance could scarcely be viewed as resulting from a tougher or more fibrous plant tissue, since fertilizing with nitrogen would be expected to do the very opposite, and make the spinach more tender, or luscious.

Here is a case where protein was protection, much perhaps, as we use special proteins to be protection for our own bodies or for ourselves. It is much like when we inject serums of various orders carrying so-called antibodies or special proteins for protection obtained from animals made immune to the specific disease in question by first having the disease. Immunities in our bodies to attacks by microbes and viruses are not obtained by introducing sugars or fats into our bloodstream. The protection is always obtained by the use of proteins.

Perhaps we shall begin to realize that insects take to plants because they are poorly nourished, growing on poorly treated soils. If we realize that insects disregard plants on soil well treated for production of proteins, perhaps we shall fertilize the soil. In so doing we shall feed ourselves better because of more proteins. In obtaining protection against insects by such methods, we shall also obtain on more fertile soils better nutrition for ourselves, and probably better protection against irregularities in our own health.

IMMUNITY AGAINST LEAF-EATING
INSECT VIA SOIL AS NUTRITION

More than three decades ago researchers in soils at the Missouri Experiment Station turned to the study of the soil as the basis of nutrition by the vision of the creative biochemistry of plants as a unique performance in which that life stratum is the only one in the biotic pyramid (a) that captures part of the sun's energy and builds it into the carbohydrates as energy and bodyfuel values for itself and for all other life forms dependent on it, and (b) that starts with a handful of ash elements from rock, soil, water, and air to create protein, the only chemical substance that carries life.

It seemed fitting that agricultural production should consider the health of the crops through proper nourishment as an important criterion of successful soil management. It certainly would impose more critical measures on that endeavor than those considering only bushels and tons of crop yields, or pounds of fattened livestock per acre, with emphasis on economic returns.

In terms of immediate value for farm practices rather than basic physiological facts, such a project would anticipate being questioned. But industry quickly reported its interest in such research. It promptly offered its support through monetary grants. Consequently the soil as nutrition was set up in its research program. It was firmly established early by the discovery that the cohabitation of the nodule-producing bacteria with the legume plants enabled those crops to use atmospheric nitrogen for increased production of proteins, but only when the plants were healthy as the result of properly balanced fertility of the soil.

Refined controls of the soil fertility established the additional fact that the available nutrients determined the so-called "fixation" or nitrogen from the air into the living protein of the legume crops. This extra supply of nitrogen, going from the soil air via microbes, root-nodules, and balanced soil fertility, was found to be a unique help by which the plants not only produced more carbohydrates as fuel foods from sunshine energy, air, and water, but also synthesized therefrom more protein supplements that promote healthy growth, natural self-protection, and immunities.

Through carefully controlled variations of plant nutrition by means of colloidal clay,[1] it was discovered that microbes and plants might be a case of either cooperation or competition. Under the soil's fuller services, including rock breakdown for required inorganic elements and the essential organic decomposition, the microbes causing nodular growths on the legume roots were not dangerous germ giving destructive and malignant tumors. They were not signs of an infectious disease. On the contrary, they indicated mutually helpful services on the parts of both plant and microbe because the soil was the properly balanced nutrition for both. The soil was the difference between healthy plants and sick ones.

Table 2

The Extent of Damage to Spinach by Leaf-Eating insects, the Crop Yields and the Concentrations of Vitamin C and Oxalate in This Vegetable According to the Increments of Nitrogen Offered by the Soil Growing it.

Soil Nitrogen m.e./Plant	Damages %Surface	Yields Gms/10 plants Fresh Wt.	Vitamin C Conc. Mgs/100 gms.	Oxalate % of D.M.
5	38.6	121.3	28.7	8.17
10	39.5	219.1	29.6	7.37
20	6.9	376.8	27.2	7.12
40	3.6	418.8	20.9	6.02

The delicate variations in the suite of the nutrient elements adsorbed on the controlled clay, as the highly dynamic soil fraction in root contact, were the means of setting the stage for many (often accidental) exhibitions of the particular nutritional conditions which were established by natural evolution and

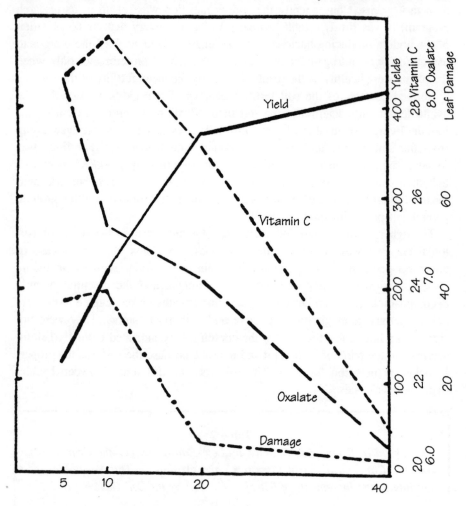

Nitrogen m.e./Plant

Figure 2. Chemical analysis of spinach crops grown with increasing amounts of nitrogen in the soils, 5, 10, 20, and 40 m.e. per plant, gave increasing yields related directly to the soil treatment. But the concentrations of vitamin C and oxalates in the plants, as well as are of leaf damage by the insects, were inversely related. (Yield = gms./10 green plants. Vitamin C = mgm./100 gms. Oxalate = % of D.M. Damage = % of leaf area.)

are required to develop a crop to its healthy climax of immunity from pests and diseases. We forgot that nature applies no poison sprays as external protection. Healthy growth through soil as proper nutrition brings internal immunity.

As a clear-cut illustration of healthy and sick plants in terms of insect pests, there was the accidental exhibition by the non-legume crops of spinach. One half was attacked by the leaf-eating thrips insects while the other half was completely immune and free of them. The damage was greatest where the soil offered the least nitrogen as plant nutrition: the immunity was greatest where more nitrogen was available. Simultaneously, there was a reduction in the severity of the attack according as the amounts of calcium available in the soil were increased.

Through doubling the amount of nitrogen given to the soil by three stages to give four rows with 5, 10, 20, and 40 milligram equivalents per plant, there was demonstrated the natural, accidental infestation by the thrips in the first two, and the complete self-protection in the second two rows. Even more significant was the additional protective effect by calcium on the infested rows as supplemental or better balanced nutrition.

Does this merely say that nature was successful in the course of evolution by simply feeding plants via soil to grow them healthy and immune to insect infestations? Does the converse not indict us? Are we not fighting infestations and infections of our garden and field crops because we do not manage the soil properly as nutrition for growing naturally healthy ones? Shall we not arrive at the all-important question and fully inquire whether crops which require artificial protection but are carried along in their unhealthy growth until harvest time can be a nutritional guarantee for healthy animals and for man consuming them? The varied vitamin C content, the accumulation of oxalates and other chemical data on the spinach were reported. Unfortunately, nature's facts reported by research have remained hidden in unopened books for one generation thus far to remind us that we learn but slowly.

At the outset here, the immunity of the plants to insects should raise some simple questions when so much has been said about "resistance." We might question the technology of "systemic insecticides," or the use of poisons fed via the roots into the plant system to destroy the insects eating the growing crop. Our vision is reluctant to picture this means as a "natural" method of providing self-protection. As an illustration, the commercially available "Meta-Systox" arouses concern since, as a poisonous sulfur-phosphorus combination, it simulates Parathion, widely known for its serious danger to humans. There are scarcely such delicate discriminations between poisons for insects and for humans when, on the cellular basis, the life processes of both are so nearly identical.

When we note, in the case of the Missouri studies, that the insects refused to eat spinach grown on the more highly nitrogenous soil, we are apt to say "the plant had developed a resistance." That term "resistance" has not been well defined. Much less does it explain just what life processes are responsible for all that is included in the term.

We forget that the insect, like any form of life, including the plant, is eating to survive. For that it must show discrimination in what it consumes, even as the plant's uptake of nutrients must vary to determine the particular survival of each species.

Should we not emphasize then the positive action of each life form going on its way to survive by nutrition for health rather than spending itself physiologically in preparing against enemies and storing substances as means to resist? Would evolution suggest loading down the individual with ammunition rather than with potentials for healthy maintenance? In viewing the soil as nutrition, we subscribe to the viewpoint of a positive health as natural for survival. The nutrition for health suggests itself as the most logical research approach, especially when soil management is the variable control.

The data, emphasizing nitrogen as the factor governing either the infestation or the immunity of the spinach crop are given, in part, in table 2. Included is data showing (a) the decreasing insect damage, (b) the increasing yields of fresh weights, (c) the decreasing concentrations of vitamin C, and (d) the decreasing concentrations of oxalate, all in accordance with more nitrogen available in the soil in which the crop grows. By presenting the data graphically (figure 2) it becomes very clear that (a) damage by insects, (b) the oxalate which makes the calcium and magnesium within the plant insoluble and indigestible, and (c) the vitamin C (ascorbic acid) all became less, while the yield became larger and crops healthier as the nitrogen in the soil was increased.

These demonstrated relations between the nitrogen and the self-protection by the plants raise some questions about the functions of the vitamin C and the oxalate within them. If the nitrogen increased growth by increased nitrogen representing healthier plants (even to the point of defense against insect attacks), then the decrease in oxalate suggests that its accumulation to higher concentrations is an index of poorer metabolism and less healthy plant growth.

Similarly, the decreasing concentration of vitamin C with healthier plants suggests that this vitamin is not contributing directly to healthier plant growth. Shall we not view its function as that of a catalyst or a "whip" to drive the chemical metabolisms of the plant more rapidly by its higher concentrations of this oxidative vitamin when the reacting nutrients become deficient or imbalanced? Considering our own foods, is a higher concentration of vitamin C a mark of higher nutritional quality? Or does it indicate less healthy growths with imbalanced chemical composition, providing poorer quality nutrition for us consuming them?

These are still unanswered questions. But they suggest that we dare not neglect much longer the research studies in soil as it determines health and immunities of all life strata if we are to stem the rising tide of degeneration of our body functions, which cannot be termed "disease" by infection.

FERTILE SOILS LESSEN INSECT INJURY

Recent soils research suggests that a more carefully balanced soil fertility may reduce plant pest and disease losses, says W.A. Albrecht, chairman of the University of Missouri Soils Department.

In many instances, the research indicates that the responsibility for disease and insect control should be with the person managing crops and their nutrition. In the past, the responsibility has been given to the chemist manufacturing poisonous drug compounds to fight vermin and disease.

Nature grew most of the present-day crops in a healthy condition before they were taken over by the human race, Albrecht points out. Generally, plants in their natural state were in a pure stand, had been growing at the same location for many years, had accumulated much of their own residues as soil organic matter, and were free from diseases and pests.

Albrecht said these facts are in direct contradiction to present-day references that state in one way or another that "it's impossible to overemphasize the importance of crop rotations to control diseases, maintain fertility, prevent erosion, and maintain soil structure."

This contradiction again suggests there is much to be learned about nature's success in growing crops with a high survival rate and without wars on diseases and pests.

Plants, as found in nature, were growing in soils where no fertility was removed. In present-day agriculture, far more fertility is removed than is returned. According to Albrecht, this is the basis for the theory that depleted soils and the consequent poor plant nutrition results in plant disease and insect attacks.

If this theory is correct, then it's logical to believe that properly balanced fertility levels would prevent such troubles and yield healthier plants.

In tests with fungus diseases of soybeans, research clearly demonstrated that attacks were highest on soils with low calcium levels. These attacks weren't evident where there were higher levels of exchangeable calcium in the soil.

More recently, improved research methods indicate that the presence or absence of leaf-eating insects varied with the levels of nitrogen and exchangeable calcium. These two fertility elements are usually connected with the production

Corn grown on soil with a low fertility level is more susceptible to insect damage while in storage than is corn grown on more fertile soil, said Albrecht. In the picture above, the small infested ear of corn was grown on a soil with a low level of fertility and was an easy mark for the grain borer. The ear at the left was grown on the same soil but with added nutrients and was undamaged except for a few borer holes where the two ears were in direct contact for a period of time. The open-pollinated ear at the right was in contact with the infested one five months but showed only one sign of borer injury.

of protein-producing legume crops.

When small supplies of nitrogen were present in the soil, thrips attacked the plants. When nitrogen levels were higher, thrips were no problem, Albrecht notes.

And, when nitrogen levels were low but calcium was plentiful, there was less damage from the leaf-eating insects than when both fertility elements were at low levels.

In still more recent work, corn grown on soils with high fertility levels was better able to withstand grain borer attacks in storage than corn grown on poorer soils. Corn, fertilized with nitrogen only, was much more susceptible to borer damage while in storage than was corn getting both phosphorus and nitrogen in the form of a soil treatment.

Ear samples from these two soil treatments were under observation for a three year period following their harvest, Albrecht said. The corn getting only the nitrogen was badly damaged while the same hybrid fertilized with both nitrogen and phosphorus was virtually undamaged.

This is good evidence that nature was able to produce crops without destruction from diseases and pests. It's also evidence that while soils may be managed to give larger yields, there's still much to learn about a soil management that nourishes plants so well that they can protect themselves from diseases and pests without poisons or medications.

Also, it raises the question of whether crops too deficient in their nutrition to grow antibiotics needed for self defense would contain enough nutrients to produce healthful livestock.

Albrecht said this poses a serious challenge for soil researchers. The challenge lies in the possibilities of formulating a balanced fertility as guarantee of less plant diseases and fewer crop pests as well as a means of producing more vegetative bulk per acre.

BREEDING OUT PLANT PROTEINS—BRINGING IN DISEASES

Much of what is apt to be considered plant breeding is mainly plant selection and then propagation of what we have selected. Such selections are usually made according to the criteria of taste, color, fruit-size, crop yield, and others. But these are not characters contributing to the plant's better survival, should it be growing in the wild and in the absence of the cultural ministrations like manuring, fertilizing, spraying and other helps for survival.

In the cross-breeding of one kind of plant with another, we select one for a certain set of characteristics and observable qualities. We select a second for another set. Then we cross these two, usually reciprocally; that is, using the first as the male on the second as the female and then vice versa in the hope for the desired combination of characteristics and qualities.

As an illustration of this procedure, a plant breeder some years ago observed that broom corn has many nodes but short internodes in its stalks, while sweet sorghum cane has long internodes but few of them. He postulated that if by crossing these two he could combine the long internodes of the sweet sorghum cane with the many of them of the broom corn, he might get a much taller stalk of sweet cane and thus increase bountifully the yield of sweet sorghum syrup per acre. The cross-bred plants resulted in long internodes and large numbers of them to give a very tall stalk. But to his sorrow, the stalk was too weak to stand up without supports as high as those common for hops. To his still greater sorrow, the stalks delivered no sugar and no sweet syrup. Cross-breeding does not necessarily combine the characteristics or properties just as we might hope.

In similar hopes much effort has gone to the cross-breeding of tomatoes to make them less susceptible, or more resistant, to the fungus attacks resulting in the wilting of the plants. This trouble is all too common in the tomato varieties giving highly desirable fruits. The wild tomato, *Lycopersicon pimpinelli-folium*, often called the "currant tomato," is not susceptible to the wilts and diseases plaguing the cultivated kinds. It is a viney, small plant producing many small fruits of about currant size stuffed with seeds but little else to invite our eating it. One might hope then, that by cross-breeding this immune, wild tomato on the choice, cultivated kinds that the resulting crossbred plants would give the regular cultivated big plants, bearing our choice delicious fruits and be immune to the wilt troubles.

Under experimental cross-breeding tests, however, such was not the good luck. The wild tomato makes proteins and packs them into seeds in large numbers in each small fruit. Apparently it also produces ample other proteins of antibiotic services to protect the plants against invasion of the fungi causing wilt. The dilution of this protective quality, by the lessening of it in the cultured variety, through the first resulting generation, did not bring on a loss of the resistance. The resulting tomato, though immune to wilt, was only a small seedy one, entirely undesirable as fruit.

When this immune first-cross was diluted more in its protein potential by a second cross with the cultured choice variety, a larger fruit resulted. As this process was continued by more cross-breeding in the same direction to develop the larger, juicy, well-meated, less seedy fruits approaching those of the choice variety, there was the loss of the immunity to the wilt. Dilution of the protein as potential in the many seeds for survival as multiplication, and also as protection against disease, gave more sugar and less seedy fruits. But there was corresponding dilution of protection against the wilt from fungi attacks.

Here is a suggestion. Our emphasis on the plant's production of carbohydrates, usually resulting in larger yields or more sugars as sweets, plus our neglect of the proteins, is inviting more plant disease because we are pushing out the element that is the plant's own protection. This protection is the proteins, or the antibiotics as compounds similar to them. The growth, the self-protection and the reproduction of plants in the wild are associated with proteins. In our selection and propagation for sweetness, larger yields, rapidly grown bulk and other desirable properties, we are pushing out the proteins and thereby opening the plant to more disease.

If this is caused by chemical changes in the plant resulting from the exhaustion of our soil fertility, should we not expect the loss of our plant's resistance to disease unless we rebuild the fertility of our soil with healthy crops in mind? Less fungi, and even fewer insects on plants grown on soils nourishing crops for more protein in them have already given an affirmative answer to that question.

CYCLES OF SOIL CHANGES IN WHITE CLOVER YEARS

Soil samples for the alternate years 1931, 1933, 1935, and 1937 were considered sufficient numbers to show the changes in the soil fertility. Since legumes like white clover require calcium (and magnesium) generously, it is significant to note the increase in the exchangeable calcium during those seven years before the one of white clover prominence.

For these four odd-numbered years the percentages of saturation by calcium of the soil's exchange capacity stepped up in the following order, 56.96, 55.09, 69.70, and 70.96, respectively. Since for better crops of legumes, like alfalfa, the experiences with soil tests correlated with the crops suggested that the calcium saturation should be brought up to 75%, it seems highly probable that this increase in the soil saturation by this dibasic nutrient might be enough to invite white clover when at the lower figure the deficient calcium was prohibiting it.

When the saturations of the soil by calcium combined with magnesium were considered, the values fluctuated between a low of 75.45% and a high of 90.85% during those years. The high value represents a soil well stocked with these two essentials for protein-producing crops like white clover and other legumes in general. Thus, we might expect the white clover to come back because the fertility improved so much in respect to calcium. Since none was applied, the incidence of the white clover suggests that either the reserve soil minerals or those recently blown in were weathered enough during the absence of the clover to bring the active fertility up again to meet the needs of the clover in this respect.

The total exchange of the soil remained constant during those years. The figures for the four samplings within the seven-year period were 19.98, 19.06, 19.20, and 19.35 milligram equivalents, respectively.

Of particular interest was the increase in the organic matter of the soil as a result of growing the bluegrass. As percent of the dry soil, the figures for the four years cited were 3.53, 3.71, 4.07, and 4.29, respectively. The total nitrogen also increased according to the following percents of the dry soil, 0.173, 0.185, 0.190, and 0.205 respectively.

One might have expected this increase in nitrogen to give it more activity in growing more bluegrass. But while this more stable organic matter of the soil was increasing, it was not becoming any the less woody or less carbonaceous to increase its rate of liberating its nitrogen, phosphorus, potassium, calcium or any other of its nutrients. This was shown by the nearly constant carbon-nitrogen ratios of it. Those values during the four sampled years cited for the seven-year period were respectively 11.85, 11.63, 12.41, and 12.15. The organic matter seemed relatively too stable for any shift to speedy decomposition without tillage. Thus the bluegrass sod was building up the more permanent organic matter. It was increasing the percentage saturation of the soil with those inorganic essentials by which the white clover would be encouraged, since as a legume it needed little nitrogen from the soil and could build its own protein-rich organic matter from atmospheric nitrogen. This organic matter, dropped back to the soil on the death of the white clover, would provide extra and active nitrogen. It would invite subsequent lush growth of the bluegrass.

This non-legume would compete with the next crop of clover for other fertility elements liberated because the added nitrogen would hasten the decomposition of the more carbonaceous soil organic matter to set them free or make them active again.

Thus we can visualize the bluegrass exhausting the soil's supply of active nitrogen and of other active fertility, and building these into more stable organic matter even with an increase in total nitrogen. This was happening while the decomposition of the reserve minerals was releasing extra calcium to saturate the soil's exchange capacity more highly with this active, essential cation. This would favor the advent and the growth of a legume crop, like white clover, which fixes atmospheric nitrogen. But this clover growth would reduce the calcium saturation to the hindrance of its own future growth. It would put extra and active nitrogen into the soil. So while eliminating itself by exhaustion of the active calcium, it would favor the bluegrass by providing the active nitrogen.

The time period required for building up the sufficiently increased calcium saturation needed for the white clover suggests itself as the time interval between the "white clover years." That interval may well include "drought" years for whatever changes in the soil or cycle they bring about.

ROOTS DON'T GO JOY-RIDING

Root behavior in the soil still challenges our understanding of it. Folks commonly speak of "deep-rooting" or "shallow-rooting" crops as if the behavior of roots spreading out in the immediate surface soil—or roots going seemingly straight down deeply into the subsoil—was a predetermined matter. Quite the opposite is the case. It is not the crop that is deep-rooted. It is the soil and the distribution of its fertility that invite that particular root behavior. Roots don't go joy-riding. They are in search of plant nutrients.

One needs only to dig into any soil under cultivation for a long time with occasional manuring to see the manifestations of the roots in relation to the manure and general fertility of it. Roots are extremely thin when they are only searching.

In the soils under an old vineyard in France, where they cultivate several times annually and plow under manure at least once each year, it was an interesting experience to dig a trench and study the roots of some more than 100 year-old vines. Cutting through the clumps of buried manure revealed a thickly packed mass of big roots.

However, the attempt to find the root leading from the base of the vine down through the soil to the manure clump was quite another matter. Only by digging carefully with a penknife or washing the manure-clumped roots out with water, could one discover the extremely fine, hair-like root connecting the mass of big roots with the grapevine for which they were serving the nourishment.

The fine root was the scout, our searching. That was the root going through the infertile soil to become the root growing in the more fertile, manured portion of the soil.

One observation of the sewer plugged by the growth of a tree root was further support of the fact that the roots searching through the soil are thin become thick once they are inside the sewer. When the sewer was dug out, there was but the one single crack in the tile connections of over 60 feet of it.

How did the root come to find that, especially when it was 30 feet from the nearest tree and about seven feet down in the soil? This is the perplexing question. Many folks answer quickly, "It was searching for water." But the root had water all along the distance of about 30 feet of its coming from the base of the tree to the break in the sewer tile. It was drinking all the time while it was going in search of nutrition. It was led to the crack in the tile by the sewage-fertilized soil around it. The more highly concentrated fertility as it approached that opening led the root right to the one opportunity in the entire sewer to get inside for some liquid fertilizer. Once inside, the root was soon growing.

The tree root inside was a mass of the size of a horse tail. It was sufficient growth to cause troubles in plumbing. Was this taking merely water? Surely not, since the intake by that root surface would certainly be at a rate far too great to be carried through the crack by the small root, and back to the tree. It was getting nutrition. This was so dilute in the sewage that even the horsetail mass of roots would absorb this no faster than the small, thin root could carry up to the tree. Outside of the sewer the root was only going. Inside of the sewer the root was really growing.

Fertilizer placement in the soil may also demonstrate the roots' small size or the slender thin roots when they are "searching" for nourishment. By putting the fertilizers down below and to the right or left sides of the seeds of beans, for example, the roots would be few, leading out in all directions. but once they entered the soil zones into which the fertilizers were placed, the roots were literally clumps of them.

In the case of sweet clover seeded on a shallow surface soil underlain by a tight acid clay subsoil, the roots scarcely entered the unfertilized and unlimed subsoil. If they did, they were too thin to be pulled out. It was necessary to dig them out most carefully to even discover them. But if the subsoil was deeply limed, then the roots of the sweet clover were a single major one of unusual thickness growing straight downward.

Root behaviors, then, are not determined by the species of the plant. They are the reaction by the plant and this soil-part of its anatomy to the fertility of the soil. It is not the crop that is deep rooting, but the soil that is such if the roots are big and numerous to mark themselves and their size out so prominently.

By the nature of the roots, then, one can judge the fertility of the soil. Thin roots tell us that there is so little of it that the roots are "searching" and finding all too little. Big, thick roots tell us that the soil is growing such sizes of them because they found plenty of fertility there.

5

PASTURES

FERTILITY REFLECTED IN FEEDING

There was a time when hay was hay, when corn was corn, when other feeds, including pasture grasses, gave animals all the nutrients necessary for growth. This was the time when minerals in the soil, and what we know as the main elements of fertility, were sufficient to supply livestock with all minerals needed for bodily functions (except sodium and chlorine supply by common salt). But years of cropping have brought changes. We notice lack of fertility mainly when crops are growing. But this same lack is just as evident in the corn crib or in the hayloft if we analyze the feeds. Indeed, it is apparent to many farmers who, by watching their livestock, know that present-day feeds lack the quality of former years.

Of course, a lack of fertility cuts yield. But it also cuts the feeding value of the crop. That in turn cuts profits—cuts profits by slower growth, by causing animals to fail to reproduce, by causing their unevenly nourished bodies to be more susceptible to disease. Increased fertility is as important from the standpoint of better quality feeds as from the standpoint of increasing yields. That may sound fantastic, but we should remember that more than 80% of Missouri's crops, at least corn, and even more of the hay, is fed at home. The grass crop also is consumed at home.

I wouldn't write this without proof. And here it is as applied to one crop— soybeans. Fifteen pounds of soybean hay grown on limed soil are equal in feeding value to 27 pounds of soybean hay grown on unlimed soil at the Missouri Experiment Station. That's increasing the value of the crop 44%. That increase might be obtained in yield—but it would be a good one.

Yet, here's the point—if we increase fertility, regardless of the method, we get not only the increase in yield but the increase in feeding value with it. We fire one shell and the results suggest we've fired both barrels. Increased fertility then has been doing more for us than we probably have thought. It's a point to consider in every green manure crop turned under, in every load of manure spread, in every pound of commercial fertilizer applied.

Barn and lot feeding of livestock have had much scientific attention. Rations and their nutritional constituents have been tested. Ration studies have emphasized the concentrates that should go with the roughages.

The roughages, and grass in particular, in the ration have not had such close study. Possibly too often we have been assuming them to be of constant feeding value. Wild animals, living wholly on roughages and given a wide range on our virgin lands, manifested deficiencies in the rations and roamed widely in their search of so-called "licks" to correct them. Any growing animal needs a great variety of nutrients, and may be suffering deficiencies in its feed because the less fertile soil cannot supply its needs on the farm to which the animal is confined.

Because of the variations in the different soils and the depletion of individual soils in their mineral supply, a wide variation in feed value of the crops produced on them may be expected. This variation occurs in the grain crops, but much more so in the forage crops. You may have heard some farmer say that alfalfa from Colorado is better feed than any he ever grew or purchased locally. We are beginning to recognize that feeds are different, even when well handled, and that they reflect differences due to no other cause than the soil.

Chemical and nutritional tests of the same crop grown on different soils are offering explanations for farmers' observations. Colorado alfalfa is richer in minerals and other nutrients than that grown on our soils in areas of higher rainfall and consequently greater degree of soil leaching. Chemical analysis has shown that sweet clover harvested from a seeding which had received limestone the previous year was rich in calcium and phosphorus, than where no limestone was applied.

Crops can make themselves only from what is offered by the soil. Animals can make themselves only from what is offered them by the crops as feed, and thus, in the final analysis, the animals reflect the fertility of the soil.

Thus, soil fertility is a factor in growing animals and should be recognized by the grower of livestock. Grass is the shortest connection between the animal and the soil, and the main feed for growing our young animals. In fact, the grass growing regions of the world gave us most of our breeds of livestock.

Experiment stations are pointing out that health and reproduction in animals are improved when there is a reduction in grain and an increase in the amount of grass consumed.

Even grass, however, is not a guarantee of success in livestock production for we hear farmers say, "Our grass hasn't the feeding value it once had." Such a remark isn't startling when in a single state like Missouri there are 175 different soils. They are different in the rocks and minerals broken down to form them, different in the humus or organic matter added to them, and different in the rate and degree to which they have lost their original supply of calcium, magnesium, phosphorus, potassium, and other essential elements which the crop must pass on to the animal to produce bone and body.

Other states have recognized feed deficiencies related to the soil. Florida has soil areas dangerously low in lime so that dairy cows in milk develop what is called "stiffs," while dry cows escape the malady.

Minnesota reports 30 counties with soils so low in phosphorus that the animals develop depraved appetites and a reduction in the calf crop.

In Idaho they have tested blood samples of the livestock and find they can thus locate the soil areas which are low in phosphorus content.

In Washington the mineral application to the soil through irrigation water is enough to make the difference between poor and successful breeding by the animals grazing on the crop.

Other studies indicate clearly that the health and vigor of the animal reflect the nutritional level of the soil.

Soybean hay grown on the limed plot of Sanborn Field, at Columbia, was richer in calcium than that grown on unlimed soil. Grasses cut for hay, and even corn fodder, produced on soil that received potash are richer in potassium than when these crops are untreated. When a crop is fertilized it may be expected to take up increased amounts of the elements offered it in larger and more soluble quantities, but the effects are more far-reaching.

All of these results indicate that a crop may be a defective feed when produced on soils of low fertility but is improved in its feeding value when the soil is treated to correct the fertility deficiencies.

WHY YOUR CATTLE BREAK THROUGH THE FENCE

In the production of livestock—like any other kind of agricultural output—we must have a margin between the costs and the sale price of the products. Unfortunately for the production of animals, more than simple controllable technologies come into play. The creation of living bodies is a biological process, and we cannot predict costs. Some costs in livestock and agriculture production as a whole are not yet catalogued. The majors among those are the contributions from the soils as chemical items we already recognize, but more significant are the creative powers for new life which as yet we know little or nothing about.

Keeping the life stream of our farm animals flowing is not as simple as keeping the assembly line in an automobile industry operating at full capacity. We still do not create livestock. It creates itself as best it can, possibly in spite of us more than because of us. Since the life stream is carried by the proteins of the feeds, and not by the carbohydrates and the fats which make up the bulk of what we offer our animals, the major problem ahead is that of growing quality proteins in our forages. Here again there is the biological process and one we do not manage except by way of the soil from which the creation of all life takes its start.

We are slowly coming around to connect the soil with the nutrition of all our life forms, whether they be microbe, plant, animal, or man. We are all too slowly coming to see soil exploitation of the past and the present as the reason why our meat, milk, cheese, eggs, and other protein foods composing the very living body, or controlling its creation, by our animals are becoming scarcer, more costly and less abundantly produced. We must look to the soil as power of creation if we are to evaluate what is ahead for any part of agriculture, especially the livestock section.

We now recognize our farm animals as very able connoisseurs of the quality of their feeds. When we accept the self-feeder, we admit that the pig will make a hog of himself quicker than we will. We can appreciate the unusual ability of our grazing cattle to assay the feed values of their forages more accurately than we can when we observe that they let the tall grass on the spot of their droppings grow taller, while they eat the short grass around it shorter. Then, too, they are always trying to get the grass on the other side of the fence, provided that other side of the fence is out on the highway or on the railroad right-of-way. They prefer such places because there the soil fertility has not been exhausted by excessive cropping.

If, in grazing, they can call the nutritional qualities so accurately in their "dumb" way of telling us, shall we not come to see that by giving attention to better treatments of our soil with limestone and other fertilizers, we can very probably build the soils for better herds? Shall we not see that when the protein-rich crops, like the legumes, are considered "hard to grow," that even plants, too, are struggling with the soil for their proteins by which their creative life stream flows to provide proteins for our animals?

Some simple observations may prompt us to ask ourselves some questions and to stimulate our thinking about the soil as it provides all that the cow would need to have coming from it, and whether a soil needs to meet only the limited requirements of being just something to be plowed, or to be manipulated by other agricultural machinery.

Forages grazed by the cow have been called "grass" and "hay." They have been something to be cut with a mower or chopped, baled, stored in the barn, and measured as tons of dry matter. More machinery to reduce the labor requirements of the farmer has been guiding the production of grasses and hays, when, in our humble opinion, that effort should be guided with the concern for more fertility in the soil for better nutrition of the cow by means of those feeds. The machinery has become primary in the agricultural picture. The cow has become secondary, if one can judge by the growing tractor population in contrast to the cow population.

When the plants we call "weeds" grow tall in the pasture while others are grazed closely and seem to be growing shorter, this is regularly considered a call for the mowing machine to fight the weeds. Instead when weeds "take" the pasture, that ought to be viewed as a case in which the cow is giving a new definition for the word "weeds." Careful observations of her behavior should raise the question in our minds whether she would suggest hormone sprays for the fight on weeds on a national scale.

She is apparently telling us that weeds are not so much a particular plant species of bad repute within the vegetable kingdom. Rather, weeds are any plants making too little of nutritional value to tempt her to eat them for it. She lets them grow taller and tolerates the degree of her own starvation required to do so. She is telling us that she does not choose her forage by the technical plant name. Instead, she chooses it by the nutritional values in the plant according as the soil fertility helps to fabricate food values in it.

Recently a case came to our attention in which a herd of beef cattle was regularly going through knee-deep bluegrass and white clover on a virgin prairie—never fertilized and never plowed—to graze out the formerly well- fertilized, abandoned cornfield of cockle burs, briars, nettles, and a host of plant species considered our worst weeds. Most of them were the kind that are under legislative bans against distribution of their seeds.

Here the cows were contradicting our plant classifications. They were disregarding what we offered as supposedly good grazing in the form of the bluegrass, and were going the greater distance to consume the plants we have always called weeds and even noxious ones. Such was their choice, though only when the weeds were growing on more fertile soil. These cows would scarcely recommend the use of hormone sprays to kill certain plant species we classify as weeds. They would recommend more fertile soils instead. Then, apparently, no plant would be called "weeds" by them.

When the cow breaks through the fence, is it her objective merely to get on the other side? A careful consideration of such cow behavior points out that she is not going from one of our fields to another one, both of which have had the fertility of the soil exhausted to a low level. Instead she is going from one of those areas of our neglected soil fertility to the railroad right-of-way, or to the public highway. She is going to where the soils are still near the virgin, fertile condition. Those soils have not been mined of their nutrient stores.

She is a judge of the feeding values, not the fattening values. She is producing offspring. She is calling for help in her struggles to keep her species multiplying. She is telling us of the shortage of inorganic substances to make bones, or of the proteins in relation to the carbohydrates. She is also telling us of the deficient qualities of the proteins in some of their parts essential for her biological processes. She will be telling us more when we learn more of the physiological processes on which she depends and which she exemplifies.

That the soil fertility made a tremendous difference in the chemical composition of a single grass species, and one considered high in the scale of nutritious grazing, was shown recently by the research of the Soil Conservation Service of the USDA, in their analysis of little bluestem of the western Gulf Region. Samples of this choice feed of the once-prevalent American bison were collected in close proximity. They showed a range in protein from 1.5 to 16%, in phosphorus from .03 to .31%, in calcium from .07 to 1.58%, and in potassium from .10 to 2.17% of the dry matter. The higher values were as much as 10 times the lower ones in the cases of protein and phosphorus, 20 times for the calcium, and 21 times in the case of the potassium.

With such wide differences, even in the ash, would you believe the cow would have survived these many years if she had not been distinguishing between such extensive variations in plant composition?

These were differences after the organic combinations influenced by, or containing, them had been destroyed by the ignition of the sample to leave only the ash. It says nothing about how widely the samples varied on the list of their organic compounds, like carbohydrates, proteins, and specific amino acids connected with the creative services these ash elements rendered.

It is significant that these widest variations occurred in calcium and potassium, which are only recently coming into consideration because of their deficiencies in the soil. Calcium has long been ammunition in the fight on soil acidity when during all that time it should have been at the head of the list of needed nutrient elements on most humid soils for the nourishment of plants, animals, and man.

The importance of the soil fertility in relation to the nutritional quality of the proteins has not yet had much attention. In this relation there seems to be much that spells deficiencies in nutrition going back more directly to the soil. In the humid soils of eastern United States we can grow corn in abundance. We are now considering a hundred bushels of this grain per acre as commonplace production. That has happened since we are growing a hybrid grain, the poor reproducing capacity of which is not recognized because it is not used as seed for the succeeding crop.

The size of the corn germ has been dwindling. Consequently the percentage of even "crude" protein in corn has been falling while at the same time the bushels per acre have been mounting. Protein production per stalk has become less and carbohydrate per stalk has become more. Capacity to help the animal make fat remains, but capacity for body growth and reproduction of the animal has fallen. We have more "go" food but less "grow" food.

When the seed of this major grass, i.e., corn, is failing in its delivery of protein within itself, shall we not expect the corresponding failure in protein delivery in the other grasses grown in the same cornbelt and harvest at near maturity as hay? Is it possible that we are moving toward a pasture system of livestock farming because only the young grass is concentrated enough or complete enough in the proteins and all the nutrient substances associated with them to nourish our animals, and keep them reproducing? Then, too, are we not compelled to depend more on growing our own proteins because the once more common protein in supplements are required for feed nearer to the points of their origin?

All these questions should bring us to connect proteins more closely with the soil under the animal rather than with only the animal itself. The cow cannot deliver proteins except as they are provided for her in the feed, save for the supplementary synthetic helps she can get from the microbial flora in her intestinal tract.

The corn plant as a producer of the more complete array of the amino acids essential for the white rat and thereby presumably for the cow, suggests its capacity for delivery of such quality of nutrition is limited to the germ of the corn grain. Complete nutritional service does not include the endosperm of that grain. One needs only to feed the whole corn grain to Guinea pigs, rabbits, or rats to see how they eat out the germ first and no more, if the grain is plentifully supplied. The complete grain is deficient in the amino acids, tryptophane, methionine, and even lysine. It is for the provision of these few deficient amino acids, then, that so-called protein supplements have always been, and must still be, supplied where the soil keeps plants from producing them.

Any other grass, like the corn plant and the grain it makes, cannot create the complete proteins required to nourish animals unless the soils growing it provide all the fertility elements and compounds the plant needs in its creative operations. Any plant is making carbohydrates when it grows. It is also making some proteins, but not necessarily these in terms of all the constituent amino acids the cattle must have to make muscle and to reproduce. It is the belief in the animal's struggle for proteins that connects good pastures with fertile soils so definitely.

The animal's ability to separate out the protein part in its feed, if given a chance, and the desperate struggle for that nutrient was clearly demonstrated by experimental trials with three pens of rabbits. They were each given 500 gram allotments of cracked corn grain which was changed to a new supply in the first pen after 25%; in the second pen after 50%; and in the third pen after 75% was consumed. By means of the rabbits' selection and consumption of more germ and less endosperm while feeding, the chemical analyses of the remnants showed that the consumed part was pushed up in protein concentration from 7.31% for the total sample to (a) 8.12 (b) 7.87 and (c) 7.53% for the consumption of (a) 25, (b) 50, and (c) 75% respectively.

More significant was the fact that the gains in weights of the animals in relation to the protein consumed tell us that the animal is selecting not for making more weight of itself. The gains as units of weight produced per unit weight of protein consumed in the corn were (a) 1.57, (b) 2.07 and (c) 2.36 respectively. These figures tell us that the rabbit does not choose to eat carbohydrates, to gain rapidly in weight, or to put on fat. His ambition for himself is quite contrary to our plans for him. Consequently if the animal instinct calls for more protein, and if this factor of the animals' nutritional wisdom has been working for better survival through better health and better reproduction when the animal roamed at large, shall we not use the animal's choice to guide us in the treatments of the soil for soil improvement? Shall we not use the animal's choice to guide our soil management for the growing of more proteins and especially those more complete in the essential amino acids by which we shall cooperate with the animals struggling to grow, rather than to fatten, themselves?

Merely more "crude" protein in the crops is also not the animal's choice, according to experiments with rabbits. A corn crop was fertilized with increasing amounts of nitrogen. The final grain crop was increased in its concentration of protein by four increments of more nitrogen fertilizers to give (a) 7.09, (b) 8.03, (c) 8.95, and (d) 10.44% of protein in the grain. The rabbits were allowed to consume but 60% of the feed allotment before it was removed. When these four samples were all before them, they ate most of (a) and (b) but decidedly less of (c) and (d). They were telling us that the rabbit does not follow us and call for more of any crude protein as we measure protein and get more of it.

Our animals are choosing according to the quality of the proteins and not just quantity of nitrogen multiplied by 6.25 which is so commonly considered the protein we talk about. We need to learn of the quality of protein as it grows or repairs body cells, protects against invasions by foreign proteins, and reproduces the species, or as the animals apparently call for it when they judge it in their feeds.

Fortunately for our beef supply, most of it resulted because the cow and her calving operation were on the western plains or the open range. It was on the unweathered, more fertile soils, that those processes took place. Now that we have exploited the soil fertility and most of our animal industry is one of fattening the animal, we are coming face to face with the need for the proteins grown to be more complete for animal reproduction. The costs in that process are reasons for artificial insemination. Failures in reproduction are blamed on microbial invasions, because they recognize a dying cow before we do. Nevertheless, with all the artificials in feeds, drugs, inoculations, or what have you, the high percentages of sterile cows, and the aborters continually hamper the profit aspect. Trouble is ahead if one can judge the trend, and the hope lies in the soil as both the provocation and prevention of most of the difficulties.

Under artificial procedures, the numerous matings required for conception, which usually succeed eventually in case of natural insemination, cause us to give up and turn the shy breeder over for slaughter. This should provoke some serious thinking. Such observations ought to raise the question whether the larger dose of semen repeatedly served by the male, in the former practice, is not a kind of successive hormone administration to bring about better ovulation and eventual conception; this may not be the case under the limited semen supply which is now used in the latter practice.

Under declining soil fertility and corresponding decline in feed values, the load of reproduction is becoming too heavy to be carried successfully during the period of gestation. If the increasing abortions and the mounting percentages of so-called "midget," calves are not studied more critically in terms of nutrition, rather than dodged by putting the blame on some aspect offering more mental escape, these losses may occur astoundingly often.

Abortion in some parts of Missouri bring about figures equally as appalling. In this latter case, the fight on some microbes—which may be only a symptom and not the cause—stimulates the search for serums from similar microbes under laboratory culture which gives an escape via what suggests a kind of blind alley. This so-called "disease" is now having the cows killed to eliminate the "disease" while it is eliminating our cows too. It is following a line of reasoning like burning down the house to get rid of bedbugs. But in the case of the "midgets" even the belief that it is a breeding problem is of no consolation. It is not limited to either beef or dairy types. Nor is it limited to any one breed in either of these groups.

A few studies of the blood chemistry of the "midgets" have found such low levels, and near absences, of some of the soil-borne essential elements to suggest the source of the trouble in some nutritional deficiencies in the cow's feed, or troubles going back to the soil for their origin. Even then, the soil-borne essential elements must do more than hitchhike from the soil through the crop, through the cow to the fetal calf. The creation of the "midget" calf suggests its irregularities traceable to irregularities in the fertility of the soil and the feed grown on it. It is leaning toward a waning faith in breeding but a waxing hope in feeding.

If our failing soil fertility is still supporting the plants as producers of carbohydrates, but is letting them down as producers of proteins, is it too much of a stretch of the imagination to see the cow being let down in calf production to the "midget" level during gestation by the low level of feeds pulled down by the deficiencies in the soil?

When the means by which body characters are transmitted from generation to generation seem so mystical to many folks, genetics as a new science is seized in hope of an explanation, especially when so much is still unknown and yet nature has done so much. The plant breeder has been hoping to breed legume forages which will "tolerate" soil acidity. He has had hopes of breeding cereals that will "tolerate" low winter temperatures, the smuts, the rusts, and hosts of other troubles suggesting themselves as manifestations of the plants' physiological inequality to the soil's limited offerings. Other aspects of the setting which involves the plant is the struggle to create the proteins by which it grows and protects itself from invasions by foreign proteins.

Only as we see feed proteins in their complete array of the quantities of amino acids balanced for body growth, for reproduction, and for protection against the invasion of foreign proteins like viruses and microbes; only as we learn more about how the cow would feed herself for offspring production rather than how we would carry her cheaply through the winter or fatten her; only as we discover the details of plant physiology by which we can know the crops which in combination will give us the complete proteins as feed; and finally, only as we know more about the soil fertility management that will undergird the plants' struggle in making proteins from the required chemical elements, can we expect to start the assembly line of the creation of livestock so that it will run in high order and without mishaps at all stages along that line.

Only as we build up the soil can we escape the fact that our proteins, which minister to better health for man and his animals, are becoming scarcer because the soil fertility for the soil's power of creation is going lower.

Rebuilding our soils is no small matter ahead as we view not only livestock production but good nutrition and good health also for ourselves.

WHAT ANIMALS CAN TEACH US ABOUT NUTRITION

In the production of feed for livestock, we have thought little about catering to the animal's taste.

The provision of each of the different feeds in quantities for balance as a good ration, then in total quantities sufficient to carry out our purpose, has been about the major thought given to animal nutrition.

We have produced hay, grain, concentrates and mineral mixtures with little regard for some of the more refined nutritional contents of each of these—that is, as these contents may be contributed or denied by the soil growing the feeds.

There is danger in thinking of a cow merely as if she were a mowing machine or a hay baler. She isn't just a labor saving method of harvesting some vegetative mass we call grass. Instead, she is a capable inspector and judge of the nutritional values of what she eats.

She cannot handle unlimited bulk to get what she needs as nourishment when that bulk is too low in concentration of nutritional elements and compounds.

Moving her from her harvesting of green rye in part of the spring to bluegrass for another period, then to lespedeza for the summer, and back to grass again in September, is merely a mowing or harvesting procedure with no thought as to the truly balanced diet in each one of those.

Should we not wonder about the "why" of it, when the cow risks her neck in going through a barbed wire fence, to get to the highway and its grass growing on soils that have not been exhausted of their fertility by continuous cropping and removal?

The cow with a yoke on her neck to keep her within the fenced area is merely mute testimony from her that she knows more about nutritional values of grasses grown on different soils than her yoke-providing master.

In Missouri, farmers trying to work with their cows so these would work better for them, have been reporting numerous cases telling us that cattle select feeds according to nutritional values.

The kinds of plants don't seem to make much difference to the cattle as long as the plants are growing on fertile soil. The term "weeds" as particular plant species is still unknown to cows.

One farmer, with considerable virgin prairie as pasture, found his beef herd going through the extensive bluegrass and white clover areas of this vegetation, to get into the weed patch consisting of the cornfield abandoned when the labor shortage of war demands took all his help away.

The unfertilized bluegrass pasture was not catering to the fastidiousness of those cows. This soil had never been given any lime or other fertilizers. They went straight through it from the water to the cornfield with its cockle burs, nettles, foxtail, etc., to eat all of what we call "weeds" and they called "feeds."

It was the fertility of the limed and phosphated cornfield, and not the species of plant, that entered into the discriminating behavior of that beef cattle herd.

In another case it was not the green plants, but their dried condition as hay that served to let the cattle demonstrate their discriminatory senses for feed quality.

In 1936, four acres of meadow area on virgin soil were given fertilizer top dressing of nitrogen, calcium and phosphorus, to the extent of no more than 600 pounds total fertilizer per acre.

This small area was part of a hundred acres of permanent meadow serving as winter feed in four haystacks, each containing the hay from 25 acres. In the summer season the hay made from the four acres was swept, with that from the rest of the 25 acre area, into the stack as one of the four in the field.

The demonstration of the cows' selection according to nutritional values occurred the following winter, when they were allowed to go into the meadow to eat the four stacks of hay. They were soon all collected around the haystack containing the hay from the four acres of fertilized soil.

This was at the opposite end of the field from the supply of salt and water. So the cows went back and forth daily from this preferred stack to the water and disregarded the other three stacks in passing them.

Even though no later soil treatments were applied on this meadow, and hay was made annually with the cows consuming it in the field, eight years later the cattle were still consuming, first, the one haystack containing, in its 25 acres of hay, the hay from the four acres given a surface application of fertilizers years before.

When farmers report that the cattle grazed out first the strip of green barley where the fertilizer drill turned over the previously fertilized drillings to double the fertilizer applied; when pigs will select the corn in the field where the soil was given treatments like lime; when they will select corn in one compartment of the self-feeder in preference to another corn in the feeder regardless of position in the feeder; when chickens discriminate between different lots of butter offered them; and when numerous other demonstrations like these by the "dumb" beasts come to our attention; shall we not believe that the animals have uncanny means of knowing more about the quality of their feed in terms of body building than in terms of fat and energy? Gradually we are realizing that it is not the carbohydrate and fat values that bring the animal choice into prominence.

Instead, it is the proteins and similar complex compounds entering into reproduction that are marked out by animal choices. These depend on the more fertile soil.

Better animals to make better foods for us will be more common, if we are guided by their choice of feeds under proper soil fertility conditions.

They judge for quality, not for quantity!

TIMOTHY HAY IN RESPONSE
TO SOIL TREATMENT

This project is in cooperation with the Department of Animal Husbandry. The effects of soil treatment usually are measured in terms of increased yield of grain, of forage, or of both. Changes in chemical composition of the crop often are given in terms of concentration as percentage, or even as total, of the different nutrient elements within the crop. Such measures of values of soil treatments have seemed inadequate in relation to the purposes for which most grains and forages are grown, namely animal nutrition. An attempt has been made to use animal growth behaviors as more responsive indexes of the changes in crop qualities in response to soil treatment.

Timothy hay was cut from plots receiving: (a) no treatment, (b) superphosphate, (c) limestone, and (d) nitrogenous fertilizers. Four lots of sheep were given a constant grain ration with these different hays as their roughage.

In terms of animal growth or average weight increase per lot the soil treatments on the timothy were of no great effect. The average gains per head during the winter were as follows: hay fertilized with nitrogen, 21.5 pounds; no treatment, 29.0; and with limestone, 31.2. There were, however, decided differences in individual weights, general thriftiness, conformation, and bone development in favor of the limestone treatments. Deficiencies and two fatalities appeared within three months in the sheep fed hay from the phosphate plot. Deficiencies occurred in about four months in the sheep fed hay from the plots receiving nitrogenous fertilizers, and the plots receiving no treatment. However, fatalities were prevented by intravenous injections of calcium gluconate. No deficiencies or deformities were evidenced in the lot of lambs fed hay grown on limed soil. This lot was uniform, active, and thrifty when dry lot feeding ceased in May.

ANIMALS CHOOSE FEED FOR QUALITY—
NOT FOR TONNAGES PER ACRE

Much research is now carried out in commercial laboratories which sell their services to industry and agriculture. Unlike laboratories of experiment stations under state-federal support, they are not a part of educational institutions. Consequently, they are not obligated to publish their findings promptly for the benefit of the public. Even though the public is welcome to follow commercial research projects, the data are prone to remain hidden in the files.

Such was true for the report of the "Cow Sense"[1] about differing nutritional qualities associated with varied harvest time of forages according to this report, the cow's choice differs from the farm manager's ideas as to the proper time to cut orchard grass for hay. The latter sets the time of cutting for high tonnage per acre. The cow sets it for high quality in the forage as nutritious feed for healthy growth.

We forget that our domestic animals, like the wild ones, discriminate between (a) different plant species growing on the same soil (prairie plant species), and (b) the same plant species growing on soils differing in fertility. By exercising such delicate choices they naturally remain healthy, grow well, and reproduce regularly. This ability of our domestic animals to choose is now being used as major help in commercial feed research.[2]

In the pattern of nature, the births of wild animals occur in the spring. This plan provides for higher survival rates for the young, as the young grass and new woody shoots of the season are maternal nursing support par excellence for mammalian offspring. But in harvesting dried grasses as hay for storage and winter use, the requisite of more yield per labor unit pushes the harvest time later into the plant's growth schedule. This economic factor violates the natural physiological requisites of good animal nutrition via quality feed as the cow would choose it naturally.

In recent feeding tests some orchard grass[1] cut for hay at three different degrees of the plant's maturity was fed to three lots of Guernsey cows. In the same field, grass was cut at (a) preheading time, (b) full bloom, and (c) mature stage. Each lot was fed under regular milking tests, with the hays as part of the ration which included also (a) ensilage, (b) grain mixture, and (c) a mat of sprouted grain grown in light to give green sprouts.

At the outset all seemed to be good rations and quantities offered were consumed completely. But very soon the cows given preheading cut hay disregarded their silage. Soon thereafter they refused the mat of sprouted grain. Then, finally, the animals reduced their intake of the grain supplement. Such was not the case with the other two lots of cattle which consumed the total ration regularly.

Here was a case of the cow telling us that quality grass (even dried) is her choice of all feeds. she prefers bulk over concentrates. she suggests that in growing forages which cows must supplement to meet their needs, we are missing the opportunity to raise healthy cows.

When these observations of the choices by the cows were correlated with the laboratory data from the chemical analyses of the hays, It was determined that by eating 30 pounds of the orchard grass hays, arranged in order of delayed cutting dates, a cow would have taken (a) digestible proteins (pounds) 3.00;

1.25 and 0.66; (b) digestible carbohydrates (pounds) 17.01;13.65 and 12.42; (c) calcium (pounds) 0.13;0.08 and 0.06 and (d) phosphorus (pounds) 0.11;0.075 and 0.004 respectively.

In terms of tonnages per acre harvested, the values set themselves up in order of (a) 80, (b) 90, and (c) 100, corresponding to the delayed date of cutting the grass for hay.

By their choice, the animals suggested (a) the amounts they needed of the four nutrients listed, and (b) the vegetative state of those nutrients within the forage plants from which high nutritional quality is to be had. The animals instinctively chooses its nutrition more wisely than the farm manager is able to select for it. In the files of animal experiences much still remains hidden from us as herdsmen.

MISSOURI SOILS AND THEIR GRAZING CROPS

The toll in terms of humus-laden surface soil already taken by erosion is leading us to use more grass crops to save our soil. These crops must go to market by way of grazing animals. We are thinking of changing from arable farming to pastoral farming in the hope that this will not only keep our better soils from eroding, but that it will also salvage income from lands already seriously depleted. In planning this change, we may well remember that our declining soil fertility as well as erosion is involved and will confront us in a pastoral agriculture as well as in an arable agriculture. The fertility of the soil must be maintained even though erosion is prevented.

It may be well to remind ourselves that the soils determine the agriculture. The climate in turn determines the soil, though we usually credit the climate directly rather than indirectly through the soil as determiner of the crops. Now that about one-half of the surface soil in about one-half of Missouri has been eroded, we are learning that we made a mistake in transplanting European habits with much tillage because of clay soils in maritime and climate with foggy precipitation to our silt soils in continental climate and torrential downpours. We now are embracing European pastoral practices because of high hopes for such in erosion prevention.

The fertility of Missouri soils cannot be disregarded in shifting to grazing crops. Declining grain yields are testifying to the declining offerings by the soil of those essential chemical items required from the soil to make grains. Grains, in general, are of relatively constant composition. When the yields decline they point out that the soil is delivering less of such things as calcium, phosphorus, nitrogen and other nutrients from which the seeds are made. When an entire plant is harvested as forage, however, we have no visible report or economical record of decline in its seed producing capacity, and therefore not any hint that the soil is failing in providing for the plants those elements that serve as their nutrients, and likewise for the nourishment and profitable growth of our animals. The composition of forages fluctuates widely as we well recognize when there may be a big plant that can scarcely produce a single seed. Herbaceousness is the result mainly of sunshine, air and water. This represents energy. The nutrient value of the forage in its passage through the young animal's digestive tract for growth effects is linked with the contributions to the plant by the soil rather than by the weather. Bone is mineral. Flesh is nitrogenous. Vegetation that is merely carbonaceous bulk does not necessarily give efficient animal growth, even though it may provide heat, energy, or fat.

Missouri soils are located in regions of higher rainfall and of higher temperature. These climatic forces have given soils naturally well exhausted of fertility elements necessary for plants. In place of lime-rich and protein-rich alfalfa as native plants readily grown, the timber crops are natural in this state. Such soils cannot be expected to produce forages that are rich in minerals, rich in protein, and rich in growth producing quality unless the deficiencies in the soil fertility are corrected by soil treatments. In putting on these soils those crops which will naturally cover the land quickly and prevent erosion, we shall be moving to those forages making their growth, less and less from the soil nutrients and hence will be of less nutritive value to the animals grazing on them. Animal deficiencies such as acetonemia, suggesting milk fever; pregnancy diseases; shy breeders; impotent males; and other irregularities recognized more readily in connection with reproduction are apparently arranging themselves in concentrations according to the soil fertility of the areas of the state. These are all manifestations to indicate that in going from arable to pastoral farming, our Missouri soils will manifest their southern location and their lower levels in agricultural possibilities without soil treatments.

Regardless of to what extent grass replaces grain or what system of farming we contemplate, it cannot continue to disregard its removal of soil fertility as different farming systems have in the past. Pasture farming will be on more dangerous grounds than arable farming has been in that it markets the soil fertility through a longer series of transformations and therefore through a greater number of hazards in going from the soil account to the bank account. In addition, it is working over the tailings of a soil mining process that has gone before. With such highly eroded and partly depleted soils as the raw material on hand, the complicated business of manufacturing animal carcasses from it for markets demanding mainly choice cuts may well put into its budget the charge against itself for the costs of soil fertility treatments if it is to give economic conditions that are better than those of the times preceding, and equal to the demands of an age clamoring for more and more improvements of them.

SPERM PRODUCTION AS A GUIDE TO
THE ADEQUACY OF A DIET FOR FARM ANIMALS

Chemical analyses have shown a difference in the protein content and other components in lespedeza hay grown on treated and untreated soil. The lespedeza hay which was grown on untreated soil analyzed 13% protein; that grown on the superphosphate treated soil, 13.8% protein; and hays on lime or lime and superphosphate treated soils, 15.9% and 15.25% protein, respectively. These hays we fed to male rabbits and the semen production studied.

Even though general thrift seemed to be maintained by all the animals, those receiving the hay grown on soil that had been treated with lime, or lime and superphosphate, produced appreciably more semen containing more sperm than did those on the hay grown on the untreated soil, or on the soil treated only with superphosphate.

Moreover, when the rabbits were shifted around so that those that had been getting the hay from the treated soil were given hay from the untreated soil, and vice versa, satisfactory sperm production continued to be associated with the hay grown on treated soil (lime or lime superphosphate).

MAKE THE GRASS GREENER
ON YOUR SIDE OF THE FENCE

That "the grass is greener on the other side of the fence" for the cow is more than bovine fancy. When the cow will risk injury from barbed wire in order to get out on the highway or into another field for the grass on the other side, surely there must be some compelling force responsible.

Perhaps you have never thought that animal instinct and soil fertility are at the basis of what may be wrongly considered as just so much "crazy cow" psychology.

Recent soil studies using animals as means of measuring soil fertility are pointing out that animal choices represent more effective gains by them, better animal health, more regular and prolific reproduction.

Lime treatments on the soil, for example, yield areas commonly selected when cows graze lespedeza in one part of a field in preference to another, as numerous farmers testify. Hogging corn down has confined itself right to the lime marking the limits within which lime was used on the soil. The corn was left untouched where no lime was used even though the hogs passed to and from through it.

Barley has been grazed out first where 200 pounds of fertilizer were applied in contrast to that with only 100 pounds. Grains, such as corn, in the self-feeder have been selected by hogs according to the soil treatment where the crop was grown. Guinea pigs have selected various grains with differences in choice according to the fertility of the soils.

Rats have selected bagged corn grain and cut into the bags according to the soil treatments, but did not select according to the different corn hybrids involved.

Experimental rats, according to the work of Dr. Curt Richter of John Hopkins Hospital, refused to eat fat when they were surgically operated so that they couldn't deliver bile for fat digestion. They refused to eat sugar if their body supply of insulin for consuming sugar was diverted. They ate more sugar, too, according to the dosage of insulin given them to encourage sugar metabolism in the body.

There is then a real physiological basis within the plant to explain possibly why the animal appetite reports different food choices according to differences in the soil.

Yes, "the other side of the fence" is sought by the animal because its better judgment as to its nourishment, and therefore its better health, better growth, and more efficient reproduction are involved. The cows usually break out on to the highway or railroad right-of-way where crops have grown annually but have not been removed.

We have been alarmed about the danger that some valuable meat or milk producer might be killed by traffic. Instead, we should be recognizing the fact that by means of the more fertile soil on the other side of the fence our efforts and the animal's time can be used more effectively in making food.

The fall season has usually been a convenient time for liming plowed soil, particularly where small grains are fall-seeded to be followed by legumes.

In case you have hesitated about liming now, why not take the efficiency of the animal production business into consideration? Soil treatments are means of increasing the animal's efficiency with no extra labor on your part. Sheep run a better wool-making business if the soil is more fertile. Animals can grow sounder bones and bigger bodies, all during the same time on the same amount of feed. Why not look to the soil fertility improvement by means of the calcium that you can put into the soil to set in motion this improved animal efficiency?

Calcium in limestone, or even in gypsum; phosphorus in the acid, or raw rock forms; potassium as salts; and all the other nutrient elements included on the soil fertility list, need our attention as soil treatments to make the grass greener on our own side of the fence as the cow really sees it.

The cows have been pleading with us, but we have turned them a deaf ear. We may well profit by using these animal assays of our soil fertility as well as by calling on the chemist for soil tests.

Soil nutrients are but 5% in the crop. The remaining 95% are air, water, and sunshine. These plentiful weather-given nutrients will not be fabricated into food for the cow, neither will the crop nor the cow manufacture food for us unless we provide an ample supply of fertility in the soil on which the big business of outdoor chemical synthesis known as agriculture depends.

We can improve the efficiency of our own efforts; we can help ourselves to more and better food when we appreciate the better soil fertility connected with the cow's psychology fully enough to make the grass greener on our side of the fence by improving the fertility of the soil located there.

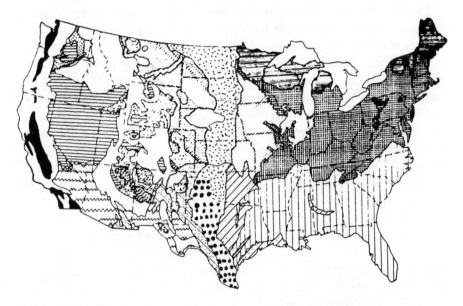

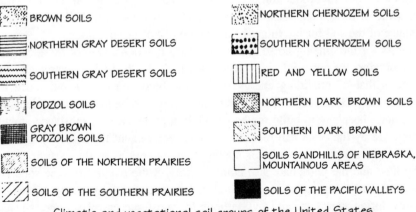

BROWN SOILS

NORTHERN GRAY DESERT SOILS

SOUTHERN GRAY DESERT SOILS

PODZOL SOILS

GRAY BROWN PODZOLIC SOILS

SOILS OF THE NORTHERN PRAIRIES

SOILS OF THE SOUTHERN PRAIRIES

NORTHERN CHERNOZEM SOILS

SOUTHERN CHERNOZEM SOILS

RED AND YELLOW SOILS

NORTHERN DARK BROWN SOILS

SOUTHERN DARK BROWN

SOILS SANDHILLS OF NEBRASKA, MOUNTAINOUS AREAS

SOILS OF THE PACIFIC VALLEYS

Climatic and vegetational soil groups of the United States.
(After Marbut, 1935)

SOUND HORSES ARE BRED ON FERTILE SOILS

Sound bones are the basis of good horses. Such bones can best be made by healthy bodies that get the necessary lime and phosphates—bone ingredients—from the soil by way of nourishing feeds. When limestone and phosphate are soil treatments to improve crops, the question often arises whether horses need dosages of these minerals directly. Our soils have much to do with the delivery of these in forages for efficiency in livestock production, more particularly for those longer-lived animals, the horses.

The fertility of the soil that includes those plant nutrients making up mainly the plant ash, is now coming in for wider general appreciation. Nitrogen as the nutrient hidden in the soil organic matter, or humus, has long been appreciated because of its scarcity and of the difficulty in replacing it. As erosion dug deeper ditches that hurried the water off in record- breaking floods and most disastrous droughts, we recently became engaged in one of the largest national action programs outside of war. We have undertaken to keep the body of the soil at home. Even there we are no longer thinking of only dams in the gully to control floods. We have gone out of the ditch and on to the upland to stop much of the rainwater where it falls by making more of it go into the soil. This unfiltered water is growing more crops as cover, more of them on the contour and otherwise, all to keep more soil at home.

The realization is about to come that it isn't so easy to grow perpetual cover crops successfully. Even before the soil started going, much of the soil fertility—the soil substance that it takes to make plants—had already gone. Not only erosion and difficulties in growing some crops are bearing testimony of this fact, but even the farm animals are reporting their troubles in deficiencies of growth and in reproduction because of our past neglect of soil fertility. These are some of the signals flashing caution and reminding us to look to the maintenance of the soil for the future health and profit of our livestock.

Grass for the grazing horse serves two body functions. One of these consists in supplying the materials by means of which the body is constructed. The other is that of providing energy to run the body machinery and move the animal and its load about. Growth demands protein to build muscle. It demands lime and phosphate to build the bones. Only a proteinaceous substance already fabricated by the plant can meet the protein requirement for horses. They cannot use the simpler elements for making their protein as in the case of plants. The lime and the phosphate also come by way of the plants from the soil. The minerals and the proteins are required in larger amounts by young horses, by mares during gestation, and by stallions for most successful service, because these are the "grow" foods and come from the soil.

Energy foods are supplied in the form of the carbohydrates and fats. It is the carbohydrates that make up the larger part in the bulk of the plants. These compounds are the main product of the plant factory as it takes carbon from the air, water from the rain source, to be combined into carbohydrates by means of the green chlorophyll in the leaves as it catches the sun's energy to do this chemical work.

Starchy grains, saccharine plant compounds, and much of the plant's fibrous structure are the horse's energy or power sources that do not come directly from the soil. They are the "go" foods that plants seemingly make from the weather, or those materials amply present without the soil. Plants won't render this energy-snatching service, however, except through the help of the 5 or 10% of their own "grow" foods that are taken by their roots from within the soil and represent the plant's mineral part, or ash.

Horses are power plants to release for our service the sun's energy stored by the plants. Horses haul their own coal when they come from the pasture. They must first be built, though, by means of the body-constructing proteins and minerals that come from the soil. Horses must "grow" first and "go" later. The soil is the foundation of our farm power plants when these are horses.

It is easy to become excessively optimistic as to what can be accomplished by the breeding of horses, now that breeding of plants has given us hybrid vigor, crosses of poultry have served for sex distinction in chicks, and certain hog crosses have given unusual growth capacities. Breeding has its possibilities, but feeding by way of soil fertility treatments needs wider consideration as to what it has done and can do for horses.

One needs only to survey the different sizes and kinds of horses in different countries and relate these to the soil to appreciate how much the soil may be in control. It is now known that soils in a region limited enough to have almost the same climate are different because of different parent rocks of their origin. As we have learned to recognize these soil differences we are likewise appreciating the influence of the soil on the development of horses.

Within small ranges of latitude and longitude reaching no great distances out from the British Isles as the center, and all within the influence of the Gulf Stream, one can go from the smallest to the largest of the horses. The Shetland pony, or the midget horse, is at home on the more rocky, less developed soils of the Islands at 60° north latitude. The Irish pony and the Welsh pony, larger but still in the pony class, are on the granitic and salty soils respectively, at 55° north. Of about corresponding size are the Russian horses on the gravelly, glacier-deposited soils of north Russia, and the Norwegian horses in the rock-bound fjords at not much different latitudes.

One needs only to go into Scotland with the greater clay content of its soils to find such active and stylish hulks of horseflesh as the Clydesdales, or east from Wales and its slates into England with its clay soils to go from ponies to the massive Shires and Suffolks. South, a bit farther, there are the heavy, close-ly-coupled Belgians. Nearby in Normandy of France on soils similar to those of England where heavy clays, heavy plows and heavy horses all go together, we find the original Normans or the Percherons of tremendous body, surplus power, and excellent disposition.

Through this small area—all within the region where woolens are the common wear—the climate is not so widely different. Yet the soils include a great variety because the different rocks in similar climate mean different minerals. Therefore the soils made from them are different. Different soils make different feeds. Different feeds mean horses differing in size, speed, conformation and disposition.

Horses differ by countries with different soils, not because horses eat the soil directly, but because even the same crop differs in its chemical composition and its service as feed for horses according as the soils are different. What, from general appearances, may look like the same soil may be decidedly different in feeding function by way of its forages. In fact, the same soil in the same place may change with time, or through neglect, enough to make its crops of greatly lowered feed value.

Soil treatments, including manures, limestone, and fertilizers, are but small additions to the soil. Yet they may alter most decidedly the nutritive value of grasses thereon. We are just beginning to appreciate different soils as they represent larger or smaller stores of reserve nutrients, and a more or less active factory with an annual output of plant nourishment available according to the crop needs through the course of the season.

Recent soil studies have reported that it is in the clay and humus, or what is spoken of as the colloidal part of the soil, that the plant nutrients are held in the adsorbed condition. This adsorbed condition means in forms available to the plant by exchange. Sandy soils in the humid region, or those with little clay, when broken out of virgin condition, are soon in crop troubles. These bring animal troubles too. Heavy soils have always been known to hold up longer. The clay is the custodian of the mobile nutrients, while the mineral particles of silt size are usually the reserve of them.

It has also been discovered that the plant gets its nourishing elements from the clay by trading hydrogen, or acidity, for them. The clay in turn, as a kind of jobber, trades the acidity to the silt particles, or the sand for their minerals of original rock nature—if they have any trading stock left. Thus, calcium, magnesium, potassium and other constituents of the plant ash are moved from the original rock form in the silt via the clay or the humus colloid to the plant roots. In the opposite direction there goes the acidity, or hydrogen, from the plant to be taken first by the clay. It is then passed on to the silt to be neutralized by its rock fragments breaking down slowly in consequence. As soils differ in clay content, or in mineral reserves of the silt and sand, necessarily the crop growth will be different. As the reserves are exhausted, naturally the kind and the quality of the vegetation will differ.

The feed quality of vegetation is largely a matter of whether it is mainly woodiness or whether it is rich in minerals, proteins and all the accessories, both known and unknown, that the better forage feeds have. Forages like the legume crops are rich in these latter respects. Can it be mere coincidence that when veterinarians so commonly recommend alfalfa hay as a remedy for vitamin deficiency they are calling on a crop that is at the top of the list for its concentration of minerals, and its heavy demands on the soil? Legumes are universally accepted as effective feed for colds and other growing animals. White clovers in the bluegrass pastures are evidence of lime delivery to the crop by the soil. Can there be any connection between the fact that clovers grow only on soils rich in lime and phosphorus, and the fact that clovers are "growth" foods? This is a question that more horsemen might raise when at last they realize that the Dutch clovers have not been seen in their pastures for so many years.

The lime and phosphate soils of France or of Tennessee and the bluegrass region of Kentucky don't mean fine horses merely because of idiosyncrasies of the people of that area. Can it be a coincidence that the winners at Churchill Down are seldom imports into that region? Such things suggest that we can connect good soils and good horses with a good likelihood that the former is the cause of the latter. Success in growing colts can not be divorced from lime, phosphate and generally good fertility in the soil. Quality of forage is more than a trademark stamped on the package. It must be grown into the goods by way of the soil.

As soils are more weathered in consequence of their location in heavier rainfalls or higher temperatures, or have been more heavily cropped, they are correspondingly more depleted of their phosphorus, lime and other minerals. As this depletion occurs, the crops on them shift to those of less growth value and mainly of a woody nature or of fuel value.

Nature has demonstrated this fact with the forests located on soils low in mineral store. Woods or timber are the last stand by vegetation against the flow of fertility to the sea. Wildlife in the forests is scant, because the minerals in the forage are scant. The pregnant timber squirrel carrying bones is no unusual observation. Well-gnawed or teeth-marked fragments of bones in the squirrels' nests would scarcely be considered as dental artifices for keeping the teeth and jaw muscles up to their maximum. Antlers in the woods disappear because of the struggle by animals to get their necessary lime and phosphorus from past animal life when they can't get these from the plant life.

Soils that are depleted, whether by nature or man, mean crops mainly of fuel value and of less help in animal growth. Not only horses, but other animals reflect these conditions in their bone troubles, teeth troubles, reproductive irregularities and alternate breeding when the more exhausted soils provide them with crops of lowered mineral contents. This principle may well be more widely applied.

Unfortunately, we are a bit late in realizing that the depletion of our soils is the reason for failure to grow white clover and good bluegrass that once were the delight of horsemen. Mechanical genius may have brought in the tractors to the extent of their elimination to bring good horses back again. Bone blemishes on horses were all too numerous in the cornbelt even before the tractor. Branded broncos from the lime-laden soils of the West were excellent examples of soundness in bone.

We didn't associate the declining store of fertility in our own soils with increasing curbs, spavins, splints and sidebones. But now that intensive cultivation by tractor, and diminishing amounts of manure and fertility going back to our soils have depleted them to the point where they won't grow cover fast enough to stop erosion, we can't bring horses back merely by economic necessity. If they are to come back economically (for back they must come under present indications) they must do so by way of better soils and fertility restoration in them.

Pasture research is going forward to give us better pastures. Much effort is being put into the search for substitute grazing crops. To date, as most horsemen will agree, there has been none found to take the place of the combination of bluegrass and Dutch clover. The clover goes out with the mineral depletion of the soil. Departure of the clover means that the bluegrass becomes less nutritious. Might it not be possible that depleted soil fertility is the reason why bluegrass quits so much earlier in the summer and doesn't begin until so much later in the fall? Can't we see the increasing need for so much supplementary summer pasture possibly connected with our neglect of putting some fertility in the form of minerals and organic matter back into the soil?

Substitute crops have come because of neglect of the soil. Now that there are more refined means of controlling the soil as it delivers nutrients to the plant, more careful study of plant composition points out that the soybean, for example, recently introduced, has a remarkable "staying power" on soils where other crops passed out.

One needs to be reminded that if soybeans are making two tons of forage where alfalfa made but a half ton, then the soil given minerals must be diluted by four times as much woody matter in the soybean forage as in the alfalfa. Soybeans have demonstrated experimentally that they may be growing to good height and yet may contain less protein and less phosphorus in the crop than was in the planted seed. They have also been shown to behave in true legume-like fashion when on soils with ample lime and phosphorus, but behaved like woody vegetation when these two were not so amply provided. Here is the explanation of why one might believe them an "acid-tolerant" crop, when in reality they shifted from a legume crop over into a timber crop.

Substitute crops are bringing on increasing numbers of disappointments for animals. Juggling of crops to maintain tonnage per acre is dropping the animals into nutritional troubles. Wide use of calcium gluconate as a remedy points toward needed attention to lime and phosphate, particularly when pregnant animals can't make it through the winter, as acetonemia and other reproductive troubles indicate. We may be trying to "rough them through the winter" but roughages from the less fertile soils are proving too rough. Substitute crops will continue to emphasize the fact that their production of mere mass is hiding the deficiency in soil fertility, which is the real need in the situation.

When the daily mineral requirements of horses are measured in fractions of an ounce while minerals as soil treatments are measured out in pounds, we readily think of mineral mixtures on the drugstore shelf as feed supplements. Even if such mixtures are helpful, this is no proof that they are a complete substitute for these applied as treatments on the soil and all that they bring with them in travelling to the horse's stomach via green or dry forage.

Recent experiments using sheep demonstrate the fact that putting the lime on the soil makes lespedeza hay, for example, a much more efficient growth producer. Liming increased the yield of the lespedeza crop by about 25%. Each pound of limed hay, however, was about 50% more efficient in terms of lamb growth resulting from consuming it. With the animals eating all the hay they could, those eating the hay given proper soil treatment made 50% more gain. Because of better crop yield, and greater growth-producing efficiency of the hay the limed acre was then about 75% more efficient in terms of increase in sheep weight.

That the lime was effective, not wholly because of the nutrient element calcium, and the phosphate not wholly because of the element phosphorus, delivered by these soil treatments is shown by these hays in digestion trials with rabbits. Contrary to expectation, the hay giving the poorer growth rate was the more completely digested. Therefore, the animal machine was handling the vegetable matter to the best of its ability.

Unfortunately, however, the unlimed hay was deficient in something to help the animal build the calcium and phosphorus into its body. These two bone-building essentials in the animals on the poorer hay were being eliminated by way of the urine just twice as fast as from the animals on the more efficient hay. These minerals were digested, but apparently the plants had not worked them into proper combination, or provided the manufactured supplement for their effective service within the body.

The mere delivery of calcium and phosphorus to the digestive tract, and a high degree of digestibility of them are apparently not enough. These essential minerals must enter into nutritional service for the plant first if they are to be of nutritional service to the animal. If these are the facts then drugstore minerals shovelled into the feedbox are not the equal in value to those put on the land as soil treatment and a help in the better output of the many complexes from the plant factory.

As the soils become poorer for certain crops and as substitutes are used, these substitute crops tend to become mere mineral haulers. Unfortunately, the minerals they deliver consist more of silica with no feed value, in place of calcium, phosphorus, and all else of nutritive value that comes with them. An unbalanced plant diet offered by the soil cannot be offset by minerals added to the vegetative bulk used as feed, and more than wheat straw would be good feed when supplemented by saltpeter, limestone and bonemeal. Synthetic diets at best leave much to be desired before they will be equal to the spring growth of the forage in bluegrass-white clover pastures on fertile soils.

Body processes of horses are not such simple performances. Neither are the processes of plant growth. When calcium in animal ash is forty times as concentrated as is the mobile calcium in the soil, and phosphorus similarly more than a hundred times, we may expect the animal to be in trouble when compelled to eat herbage getting little of these essentials from the soil. Animals know their forages so well that even a blind horse, according to Doctor Dodds of Ohio State University, will graze to the line of the soil treatments represented by only a few hundred pounds of fertilizer.

We might then expect that the thousands of pounds of fertility hauled off through years of farming are a decided disturbance in animal behaviors. In place of going to the drugstore for mineral supplements, it would seem better to let the animals make their selection via plants from a liberal variety of them in the form of fertilizer put back on the soil. Animal production is not wholly a matter of shortcuts and economics, but a most discerning cooperative effort on our part in the complex performances of nature.

Fortunately, the cornbelt and much more of the United States were blessed with good soils, particularly good for horses as pre-tractor days demonstrated. They will be good soils for horses again if we treat our pastures with the proper mineral fertilizers to restore white clover, the bluegrass fertilizing legume. Fertility depletion during the youth period of Americanism toward our soil need not prohibit our handling it from this day forth with the maturer judgment of American adulthood apparently about to arrive. We can hold our soils at the present level, and even build back.

Horses can help us in this program and guide it by their help in their more refined assay of the mineral nutrient levels. Short-lived animals suffering early nutritional irregularities reveal soil troubles quickly. Meat producers like sheep are more responsive than hogs, of which the fat constitutes most of their bulk and hides the trouble more readily. When it comes to the long-lived horses their troubles more slowly accumulate and longer remain hidden as minor defects. For these hidden faults, prevention is better than antidotes or cures. For prevention there is no better means than good feed of animal choice and collection from fertile soils. For horses, as for humans, the way to be healthy is to be well-fed.

Just as good horses are supported by fertile soils, likewise good horses in turn support the fertility of the soil. Horse power is merely the sun's energy released on the farm right where it was collected by means of the soil minerals or soil fertility. None of this needs to be exported from the farm to pay for horse power as is done for liquid fuel and lubrication. In using horse power, the soil fertility is merely in rotation from the field to the barn and back to the soil. As the amount in this cycle of rotation becomes larger, the yields of crops on the farm go up.

Horses as additions to other livestock can be significant additions to our efforts in soil conservation in a larger way. As more of the minerals mobilized out of the soil into organic matter combination go back to the soil, these are more nearly good growth food for the soil bacteria. This puts "life" into the soil, which with more grass pastures will win its way back toward virgin condition. It was—and will be—on such mineral-rich and humus-rich soils that our lands will do much to conserve themselves. It will be because they can take the rainwater and because they have the fertility to grow their own cover to prevent erosion. It was—and will be—on such soils that the production of sound horses need not alarm us much about the necessity of supplying them extra minerals.

FEEDING THE ANIMALS BY TREATING SOIL

The wider appreciation of the soil is now generally recognized, (under the impetus of extensive federal appropriations to make it so). Many are the confessions at the soil conservation altar of our failure to hold on to our greatest natural resource, the soil itself, now that in Missouri the equivalent of one-half of the top soil in one-half of the state has already moved toward the Gulf of Mexico. We are fully agreed that the very body of the soil has been washing away. We are also agreed that something must be done about it, (if we get enough cash for doing it). Quite unnoticed, unheralded and unsung, however, is the real loss from the soil, namely the loss of its fertility. Who is championing this great cause, by which the very productivity of the soil has been going down, even if we keep the soil body at home? Here is the real soil decline, and yet few are they, either in private or public, who are striking significant blows to lessen it. It is this item that we are aiming to bring to your attention by pointing out that even the dumb brutes of the field have been recognizing it, even if their masters have not.

What is soil fertility, you may be asking. It is that part of the soil that serves as part of the plant structure or the ash portion. The ash of a plant, you say, is scarcely more than 10% of such a good mineral-carrying and bone-building crop like alfalfa. So with four tons of alfalfa, but 80 pounds of ash or soil are taken from the land, and this is but a mite if an acre seven inches deep weighs two million pounds, you may counter in remark. Have you ever reminded yourself that of the two million pounds of soil we call an acre, only 13% of it, under the most fertile conditions, represents the entire stock of plant nutrient possibility, if all of this one-eighth could be made into such a form that a plant could use it? Remind yourself further that this 13% has come down to us after being able to resist weathering for not only centuries, but for geological eras. It is from this 13% that has been tried through ages and found resistant to weathering that we ask our plants to satisfy their ash needs through contact during a single short growing season.

How then can a plant feed on the soil, or rather, let us remind ourselves that a plant doesn't take it from the soil but rather the soil gives it to the plant. Let us look into the soil to remind ourselves that it consists of sand, silt and clay with some 5% organic matter. The sand fraction is still of larger rock particles which have held out against weathering and therefore consists of quartz. As a fertility contributor, this has a zero value. The silt in most soils is also highly quartz and has 30%, or less, of "other than quartz." This 30% is made up mainly of such minerals which have little plant nutrient value. Hence, the silt, like the quartz, is thus of almost no nutrient value, even in terms of content, to say nothing of no value in terms of solubility, or "availability." Thus, the sand and the silt have a "site" value, a place where plant food might be put, or through which plant roots might forage in search of plant nutrients. By elimination then the clay fraction and the organic matter portions of the soil must be the main source of what will nourish the plants. These represent not inexhaustible providers of fertility, but merely temporary stores of soil fertility en route to solubility and to the ocean.

The organic matter fraction is merely the soil nutrients that were made soluble previously and caught up by plants to be held over in organic combination, and to be set free again as soluble minerals for repeated plant use when the organic matter is digested from them by bacteria in their struggle for energy. We have not been fully cognizant of the significance of organic matter decay as a liberator of mineral nutrients. We have recognized its value for nitrogen, but if you will connect the observation of legume nodule clusters in clumps of decayed organic matter in acid soils, and the general observation when a farmer says "fertile soils are not so sensitive to acidity," you will have basis for reasoning that organic matter decay liberates calcium, magnesium, and potassium or the bases that legumes take from the soil. Organic matter is thus an item in fertility because of its content in minerals, or bases, as well as of nitrogen, and also as the colloidal absorber of nutrients released by the slow process of mineral breakdown.

The clay fraction is the real source of nutrients in the soils outside of the organic matter. Not, however, because of rock or mineral crystal breakdown by the clay, but because clay "adsorbs" nutrients. This means that the clay can take up the nutrients from solution, as you know its filtering capacity, its power to take odors, gases, or other items because of its tremendous surface on which these are held. The clays give up to the plants these items they have adsorbed. They can't give up what they haven't adsorbed or what hasn't been taken by the plants from the clay but only as the plant exchanges hydrogen for these nutrients and thus leaves the soil loaded with hydrogen or acid.

Soil acidity then is merely a case of nutrient deficiency. Continued cropping is thus bringing more acidity, and exhausting more and more the stock of exchangeable nutrients on the clay. The degree of soil acidity is merely a measure of the degree of fertility exhaustion, for as more hydrogen has gone into the soil to make it more acid, more of the nutrients, calcium, magnesium, potassium and other positive ions must have gone out to let hydrogen have their places.

What do plants need to make their growth and what fertility items from the soil are declining seriously, is the next question. Fortunately the problem of nutrients is as simple as 14 elements. These include the following: carbon, hydrogen, oxygen and nitrogen, coming from the air and water. Boron, manganese, copper and zinc, coming from the soil, but used in such small amounts that they are deficient under only rarer soil conditions—calcium, phosphorus, potassium, magnesium, sulfur and iron, coming from the soil and needing attention as to supply.

Our soil fertility has been declining regularly but our American agricultural youthfulness has not yet brought us to the consciousness of our prodigality. We have marketed our fertility abroad until we require now almost an army to guard the gold taken in return for it. Chemical data show much concerning soil depletion, but instead of citing such let me hurriedly give you about six or seven bold indications of our soil fertility decline that are strong evidences which we have almost totally disregarded, though of common knowledge. The decrease in the 10 or 15% of the earth's crust that offers sustenance to plants has dwindled to the point where its seriousness is now shown by the following:

1. The erosion of the soil itself.

2. Declining grain yields, or no increase acre yield average in Missouri in spite of breeding and selection of crops.

3. Shift in nurse crops for legumes.

4. Increasing legume crop failures and shifts in kind of legume crop.

5. Greater weather hazards.

6. Incidence of particular crops on abandoned land or acceptance of crops with lower protein producing value; and increase in deficiency diseases.

Perhaps you have given little thought to the natural incidence of weeds or crops on abandoned lands, or to our importation of certain kinds of new crops as they are suggestions regarding declining soil fertility. But let us remind ourselves of the following simple facts. First, the greater mass of a plant is carbohydrate, or of carbonaceous composition. As plants are more desirable for forage feed or human food they are high in their percentage content of proteins, minerals, or vitamins. They are better if selected when young because of the dominance then of these components within the plant. Young plants are relatively rich in these, since they get at the soil's store in early spring, or when the soil's supply of accumulated nutrients is higher because of the winter and of the early spring microbiological activity. Later in the season when this supply is gone they do not run their factory on what the soil gives, or on the calcium, potassium, phosphorus, that make protein, but are limited to what the air, water and sunshine offer. Thus, they become dominantly carbohydrate factories.

We have been blaming climate for this difference, or the social order situation in the South. But within our own constant climate we are shifting our crops from those which are proteinaceous producers to those that are carbonaceous producers. Incoming weeds demonstrate this same fact. We can now demonstrate these facts in experimental conditions where we can shift legumes from being proteinaceous factories to becoming carbonaceous factories by the lowered level of calcium and phosphorus. This has been happening on our farms. Our legumes have become wood makers instead of feed makers, accordingly as the soil fertility in calcium and phosphorus has been going down. Non-legumes fit into the same category and on our calcium and phosphorus deficient soils our hays from the grasses have been almost in the excelsior class of the horse that died just when he had learned to eat it.

Thus, with the decline and shortages in soil fertility, particularly calcium and phosphorus, our crops have still been labeled as they once were when grown on soils of higher organic matter and higher clay saturation of bases of virgin state, but their performance as factories has been going down toward the level of cotton fiber or wood pulp of the less fertile soils of the South.

While we have not been appreciating this, our animals have. As we shift our farming under soil conservation guidance from grain to forage that is measured by the top, these changes in fertility are in danger of being less appreciated. We are already face to face with the simple problem of treating the soil to calcium and phosphorus so as to enable the plants to run their protein, mineral and vitamin factories and thus feed the animals at respectable efficiency. If we go more to grass it is all the more essential that we appreciate the declining soil fertility to maintain our livestock on an efficient basis or before we let them down to the 30% calf crop, for example on the less fertile soils of the South. Because cattle on the forages grown on the unleached soils of the West can "rustle for themselves" or "rough it" through the winter as we hear it said, we will need to remind ourselves of the differences in fertility offerings in the grasses of the semi-arid conditions of the West in contrast to those of the humid cornbelt and South. Already some of our livestock is rustling to the limit, or roughing itself almost to death in attempting to go through the winter on our forages and carry the additional body load of reproduction. Should we disregard all the other evidences of declining soil fertility, surely in the face of these animal aspects we can see more clearly that our declining soil fertility is bringing us to realize that we must in some measure feed our animals by treating the soil.

6

ANIMALS

PASTURES AND SOILS

Our grazing animals are gradually compelling us to credit them with an uncanny ability to judge the nutritional qualities of the forages they take. We are coming to realize also that the higher nutritional qualities of the forages are determined by the higher fertility of the soils growing them. We are looking at pastures then, not so much in terms of their acres, but according to the fertility of their soils and thereby the nutritional quality of their forages.

We can appreciate the unusual ability of our grazing cattle to assay the feed values of their forages more accurately than we can when we observe that they let the tall grass on the spot of their droppings grow taller, while they eat the short grass around it shorter. Then, too, they are always trying to get the grass on the other side of the fence, provided the other side of the fence is out on the highway or on the railroad right-of-way. They prefer such places because there the soil fertility has not been exhausted by excessive cropping. If, in grazing, they can "call" the nutritional qualities so accurately in their "dumb" way of telling us, shall we not come to see that by giving attention to better treatments of our soil with limestone and other fertilizers, we can very probably build the soils for better results with our herds?

"But just how shall we know what a soil must contain or what treatment should go on our soils in order to grow better cattle," you may already be asking.

Perhaps you have already had some answers to that question. Suppose we look at the fertility outlay in our virgin soils or what the soils contained in the days before they were tilled and cropped. Perhaps some of them can offer suggestions.

You will recall that westward from the Mississippi River, or even for some distance east of it, there were the prairie grasses on the prairie soils. These lands were not growing herds of domestic cattle when the pioneer first ventured into that region.

The prairies were, however, producing those big-boned, well-muscled hulks of animal flesh of similar feeding habits, namely the bison. These animals were meat and hides for the Indians. They were entirely grass fed. They were grown without any purchased protein supplements and, apparently, must have found their native forage of the proper nutritive ratio. One needs to ask why buffaloes had chosen, and stayed on, a particular soil area. These meat producers wandered north and south over almost the same soil belt as that where we now grow hard wheat. This is the crop producing a grain which is recognized as good food, particularly in whole wheat bread, and a plant of which the young growth makes good pastures for lots of Herefords today.

There on the plains was the soil that was naturally built for better herds. There were the soil contents as nutrients that can be put through the modern chemical tests and bioassays to give us a good pattern soil for beef production. There was the fertility condition toward which we can build up the other soils by the necessary soil treatments. That is the kind of soil that we want to duplicate when we use lime, nitrogen, phosphate, potash, magnesium, and the less prominent, or "trace," elements as treatments to improve our soils.

Prairie and plains soils were fertile in calcium or lime. Streaks of this were found down in these soils within a few feet of the surface. They contained reserves of minerals supplying other nutrients. Those soils were acid enough to breakdown and make these nutrients available to plants. Legumes were growing naturally and thereby providing feeds that were growing young animals rather than merely fattening old ones. Soils were not as acid as those farther east. On the more eastern soils today we know that the difficulty in growing the mineral-rich, protein-producing crops for more efficient feeds is due to the absence of calcium and other fertility that went out when the high rainfall and heavy cropping put acidity there in exchange for them.

The prairie and plains soils were well supplied with nitrogen, because when once it was taken from the air by the legumes to construct their vegetative bulk and then either grazed off by the bison or left in the grass, this fertility element went back in humus form on the death of both. The soils' contents of phosphorus were assembled from the rock minerals by the searching plant roots. This element was left in bountiful quantities in the surface soil layer in the organic and highly available form by this cycle of growth, death and decay in place.

Large amounts of potash were also active in soils so heavily stocked with organic matter going down so deeply. Being in regions of moderate rainfalls and periodic droughts, and being traversed by rivers hauling unweathered mineral sediments eastward from the arid region to be blown as deposits over them, these soils had regular additions of inorganic nutrients. Surely then, the chances were running high here that even the "trace" elements, like manganese, cobalt, copper, iodine and others would be present in the soil and the vegetation. At any rate, all the nutrients were there in the quantities and in the territories sufficient for bison. Those soil areas provide the fertility pattern toward which we can build our other soils today by our soil treatments.

While the prairie and plains soils of our mid-continent still support herds today, they do less effectively according to the proportion in which we are depleting the soil fertility by moving it off the farm in the grain and the livestock. Our declining soil fertility has been pushing beef cattle westward. This movement, some might say, is due to economics. But when the "hard" wheat, or high-protein wheat, is also marching westward with low-protein wheat following, it may not be erroneous to see the animals as the protein manufacturing part of agriculture, going west when the protein-producing plants go out there, because the needs for lime and other fertilizers for them on our soils are constantly growing more serious.

Perhaps you will grant that the exhaustion of soil fertility caused the movement of the big beef cattle market from the East to Chicago years ago and its movement lately from there to Kansas City. When we remember that the soils of the eastern and southeastern United States were originally low in their fertility because this was washed out by high rainfalls, we can readily understand why our beef animals can reproduce and be grown more efficiently after those soils are treated with limestone and other fertilizers to correct the fertility deficiencies. Such treatments make better herds. These result because soil treatments make the once-forested soils and their forages more nearly like those of the original prairies and plains, where protein production as well as carbohydrate production was more readily possible.

If the soils were originally forested, they were then already relatively low in fertility even before we cleared them. They were providing potash, which helps the plants use the air and rain water to build carbohydrates, that is sugar, starch, and cellulose to make their wood. But potash is rather badly out of balance with other less prevalent elements for the plant diet that, would enable the plants to make protein and big seed crops. The diet for the vegetation was already too low in calcium or lime, which is the soil requisite that we in the eastern United States associate with protein production, for example, by legumes and with good bone development and body growth in young animals. The plants' diet is out of balance for the manufacture of very much protein when that diet is growing only forests naturally. Soils in forested areas are deficient in one or more of the essential inorganic elements, and thereby deficient in furnishing a complete list of the fertility elements.

That we can apply lime, phosphate and other fertilizers and then get response by the animals has been shown: (a) by their choices in grazing of the treated soils first; (b) by their greater gains or young animal growth on forages from fertilized soils; and (c) by better reproduction on treated soils.

On many so-called "acid" soils one needs only to apply limestone, gypsum, old plaster, cement, acetylene waste or other calcium fertilizers as a streak across the pasture to see how the cattle will keep this grazed closely in contrast to the taller grass in the rest of the pasture. Such demonstrations have been numerous by the Missouri farmers, already well given to feeding their animals better by treating their soils.

Areas given phosphates and more complete fertilizers carrying nitrogen, phosphorus and potash are also similarly selected first for grazing. Barley to be grazed was fertilized at a double rate where, by drilling out the corners, the turn was made over parts already seeded and fertilized. The doubly treated spots were taken first by the Herefords when put into the field to graze it. The cattle select not only the first crop after the soil treatment, but they keep selecting one of the four haystacks each year regularly for eight years after only one single treatment was put on the surface of this virgin prairie soil. This occurred when less then five acres of the produce of the 25 in the haystack were grown on the fertilized part of the hundred acres.

Almost any soil treatment that makes up a deficiency in the fertility of the soil to balance the plants' needs more completely, seems to make forage of greater preference by the grazing livestock. When the cattle exercised a choice of forage in the field, can we not expect their growth rate to be higher from it, just as we know it is for the hog exercising its choice at the self-feeder?

That building better soils also builds better herds was demonstrated forcefully by one of our Missouri Hereford breeders. Several years ago, as a beginner in growing whitefaces on the acid soils on his farm, he decided to build up the soil as well as his herd. He has been testing his soils, liming, plowing down fertilizers, using legumes, and, while he has been building his soil deeper, he has been highly satisfied to see that he was building the quality, health, and reproductive powers of his herd much higher. His herd is much improved over that of the nearby herd from which his was established, and which is on the originally similar soil type that has not been built up as his had.

Soil treatments using phosphate are proving to us that this element is put into the soil to come up through the crops, and to bring with it some synthetic effects from the plants which, like those from calcium, cannot be duplicated by putting inorganic forms of these into the mineral feed box.

Other elements are doing likewise. Copper put on the soil as only a few pounds per acre in south Australia has "cured" sheep troubles known and feared for a long time. Perhaps manganese, used as a cure for perosis in chickens when added along with these other "trace" elements as soil treatments, may soon demonstrate its value in this way also. If it, along with these others, should reduce a baffling disease like brucellosis as may be suggested from some effects thereby on undulant fever, then better nutrition resulting in particular blood and body proteins, and coming via the soil becomes protection against a baffling disease, and one supposedly transmitted from the animals to the man. This protection would be obtained by humans according to the degree in which the animal is first protected by specific fertility additions as soil treatments.

The importance of the soil fertility in relation to the nutritional quality of the proteins has not yet had much attention. In this relation there seems to be much that spells deficiencies in nutrition going back more directly to the soil. In the humid soils of eastern United States we can grow corn in abundance. We are now considering a hundred bushels of this grain per acre as commonplace production. That has happened since we are growing a hybrid grain, the poor reproducing capacity of which is not recognized because it is not used as seed for the succeeding crop. The size of the corn germ has been dwindling. Consequently the percentage of even "crude" protein in corn has been falling while at the same time the bushels per acre have been mounting. Protein production per stalk has become less and carbohydrate per stalk has become more. Capacity to help the animal make fat remains, but capacity for body growth and reproduction of the animal has fallen. We have more "go" food but less "grow" food.

When the seed of this major grass, i.e. corn, is failing in its delivery of protein within itself, shall we not expect the corresponding failure in protein delivery in the other grasses grown in the same Cornbelt and harvested at near maturity as hay? Is it possible that we are moving toward a pasture system of livestock farming because only the young grass is concentrated enough or complete enough in the proteins and all the nutrient substances associated with them to nourish our animals, and keep them reproducing? Then, too, are we not compelled to depend more on growing our own proteins because the once more common protein supplements are required for feed nearer to the points of their origin?

All these questions should bring us to connect proteins more closely with the soil under the animal rather than with only the animal itself. The cow cannot deliver proteins except as they are provided for her in the feed, save for the supplementary synthetic helps she can get from the microbial flora in her intestinal tract. The corn plant as a producer of the more complete array of the amino acids essential for the white rat and thereby presumably for the cow, suggests its capacity for delivery of such quality of nutrition limited to the germ of the corn grain. Complete nutritional service does not include the endosperm of that grain. One needs only to feed the whole corn grain to Guinea pigs or rats to see how they eat out the germ first and no more, if the grain is plentifully supplied. The complete grain is deficient in the amino acids, tryptophane, methionine and even lysine. It is for the provision of these few deficient amino acids, then, that so-called protein supplements have always been, and must still be, supplied where the soil keeps plants from producing them.

Any other grass, like the corn plant and the grain it makes, cannot create the complete proteins required to nourish animals unless the soils growing it provide all the fertility elements and compounds the plant needs in its creative operations. Any plant is making carbohydrates when it grows. It is also making some proteins, but not necessarily these in terms of all the constituent amino acids the cattle must have to make muscle and to reproduce. It is the relief in the animal's struggle for proteins that connects good pastures with fertile soils so definitely.

Two world wars that were fought under the slogan that "Food Will Win the War and Write the Peace," ought to encourage our inventory of the soil resources that were the food resources by which one group of the fighting nations became the victors while another became the vanquished. We may well look to the soil fertility supplies by which the Three Great Powers emerged in the category of that distinction and only by which they will stay there.

One needs to look at the soil map of the world and to remember that proteins of high food value as found in hard wheat, beef, and mutton, for example, are the products of soils that are only moderately weathered. Such soils and such protein products, then, must occur under moderate rainfalls and in the temperate zone. Such soils with extensive areas of hard wheat and animal herds in large numbers occur in the mid-continental United States. Likewise there are similar extensive areas in the Soviet Republic. It is these soil fertility resources in terms of protein production that give strong suggestions why these two nations are listed among the Great Powers. As for England in this category with them, the British Isles do not have extensive areas of soils that produce hard wheat. But when Canadian soils represent high protein-producing powers, and corresponding soils are extensive in Australia and South Africa—all parts of the

British Empire—there is ample suggestion that ships on the sea represent the strength of this third one of the Three Great Powers.

The strength of any nation—in what is too readily considered as a political strength—depends on high levels of fertility of the soils that represent protein production as food. The weak powers, under the analysis for their soil resources, all reflect very clearly their insufficiency as producers of food proteins. It is in terms of soil fertility resources and not of international politics that the world must be inventoried if we are to understand and solve the international food problem. We must realize that it is very acutely a protein problem rather than one of only calories.

Seemingly we are still nomadic in our hopes and in our thinking about our future food supplies. We are delayed in realizing that about all the land areas of significant protein power have been taken over and put into production. We are still more delayed in appreciating the problem of maintaining in the future the capacity to produce protein where such was a simple matter in the past. We need careful inventories of the fertility resources in our soils, and of the supplies of minerals that can serve as fertilizers in soil fertility restoration. Those of us living in cities, those managing big industries, and all in the congested food-consuming rather than food-producing centers need to understand and appreciate the rate at which our soils are being exploited and not rebuilt. All of us need to aid and encourage soil restoration in terms of those nutrient elements serving in the struggle for protein (a) in the life of the microbe in the soil, (b) in the life of the crops in the field, (c) in the life of the animals, and (d) in our own human lives. We need to realize that T-bone steaks are not grown on city pavements, but only where the fertility of the soil keeps the assembly lines filled with the raw materials on which all agricultural production, and thereby food production, depends.

Good beef as good protein for excellent nutrition of ourselves will come from a grass agriculture only when the pastures are on fertile soils.

SOIL SCIENCE LOOKS TO THE COW

The practice of fattening beef cattle (as well as raising them) has now moved out to the West, too. It has come out to where feeder cattle were grown to be shipped east for that treatment before their slaughter. Now that the fattening is threatening to supplement but not replace the growing of cattle in the West, we ought to recognize the simple, natural fact that the raising of cattle in the Midwest and on the plains and the fattening of them in the cornbelt, were part of the pattern of agricultural practices premised on the fertility pattern of the soil. Fitting those practices wisely into the fertility pattern was a natural accident. It can scarcely be credited to our early recognition of how powerful the forces of creation within the soil really are. We are coming to recognize the power of soil fertility, even to its differentiation between the possibilities of growing young animals and of merely hanging fat on those already near grown.

It certainly was not our knowledge of the different soils of the country which marked out the limited soil areas where we can put proteins into the muscles and minerals into the bones of our meat animals effectively. We have not seen the less-weathered, mineral-rich soils in the mid-continent as the reason why cattle grow themselves there. Nor have we seen the highly-weathered, mineral-poor soils of the East as reasons for feeds that only fatten there, and even then only when some mineral and protein supplements brought in from some other soils are wisely supplied. We talk much about better land use and much more about farming according to land capabilities. But capabilities on the part of the soil as its fertility gives feeds complete to grow animal life without calling in all kinds of supplements from the grain markets, the drug stores, and veterinary supply houses, have not yet become common in conversation. We see our problem of supplements as something to be purchased. Instead, it might well be viewed as something we need not purchase, if we were to test our soils for their deficiencies and to learn what soil treatments will help to grow complete feeds.

It was Horace Greeley, that famous New York editor, who kept suggesting, "Go West, young man." But he did not know that such was excellent advice in his day for one hoping to grow beef cattle. Greeley might well have supplemented his suggestion to the Junior Cattlemen of that time by saying "Go West to where the soil will grow good cattle, since there the Creator, himself, had a big herd of cattle-like bison or buffalo." That was already a case of the cow ahead of the plow. It was in reality the cow instead of the plow with continuous cover on the soil and the minimum of its erosion. It was grass agriculture. It was grass, though, with a reputation as nutritious feed carrying growth power more than only fattening power per mouthful. It was help for every little buffalo to become a big one, and for every big one to make a lot more little ones.

Grass in New England—since corn is a grass—didn't deliver such growth power to animals, because it couldn't get it from the soil there. Even the Indian had to put supplementary plant foods under every corn hill in the form of a fish as fertilizer. Going west was a move away from corn requiring soil fertility supplements to grow its yield of grain, and this, in turn, demanding protein supplements for growing the meat animals. It was a move to less rainfall and to soils not so highly leached out that wood is all they can create. It was a move to soils that are wind-mixed from all kinds of fertility material blown in from everywhere; and to great depths of such fertility supporting the protein-rich, nitrogen-fixing legumes which were thriving naturally to feed growing young animals rather than fattening mature and castrated ones. Going west was an escape from the need to find what supplementary feeds must be provided for making the proteins for growing the animals.

In the midwest cattle originally did not take to eating limestone from the "mineral box." Yet an area of no greater length from east to west than the extent of Kansas was enough to let the cow tell us that she would take limestone as her "medicine as supplements to her hay grown in western or mid-Kansas. When she stayed in the eastern part of that state she took to the minerals and the drugs. On going west she dropped the habit.

The cow, too, might seem to have been following the advice of Mr. Greeley. But she was not given a chance to guide our agriculture in that direction to the same degree as she exercised her leadership for the nomad and for primitive agriculture when the cow went ahead of the plow. Left to herself, she might have gone ahead of the plow to our good advantage in more recent times. She might well go ahead now to make biochemical assays of the forages and thereby to select the soils promising good nutrition and good health for herself and for her owner at the same time. The cow's coming west was an escape from the problem of protein supplements. It was a move away from lands where legumes are "hard to grow," or where some legumes, like the newly introduced soybean, were considered "a hay crop but not a seed crop." It was a move to grasses requiring no fish-fertilizer to guarantee their own reproduction via seed. It was grasses guaranteeing the reproduction of the buffalo, of the beef cattle, and of the "hard" wheat—all with no more attention than being turned out "to rustle for themselves."

But just as the soils to the east of the mid-continent did not permit the bison to survive there, so the soils also west from there excluded him. We are gradually recognizing the buffalo pattern as the pattern of soil fertility potential for high-protein wheat and high-protein meat in our beef and our lamb. We are recognizing the soil as responsible for the low-protein and high-fat meat in the pork of the cornbelt. It is unfortunate for us that the cows on the range have taken care of themselves so well by their own capacity to choose grazing according to better soils. As a consequence of their good soil sense we have learned all too little and all too late about the fertility of the choice soils under the cow's cataloging of their capabilities. Of course, we are testing soils and are fertilizing them for more bushels of corn per acre, and for more fat on the steer. However, we are not yet fertilizing for more calves delivered regularly and for those with more muscle, less fat, and more stamina, as calf crops of 60% testify.

Now that we, and the cows along with us, have overrun the better protein-producing soils in our westward dash, the decline in the animal's reproductive potential—shown by less ready conception, less vigorous calves, and more infectious troubles associated with calving—is suggesting that even the soils of the former buffalo area are being seriously mined of their fertility. It suggests the soil's decline to the degree where feed supplements for the cattle and fertility supplements to the soil for the grasses are a problem calling for prompt solution, if beef is to be grown even with effort where once it grew readily with little work by the cattlemen. The problem of supplements has come west because we as soil miners moved our agriculture in that direction.

Proteins are the complex substances we, and also our animals, must eat if the bodies of both are to grow, to protect themselves against ailments in health and to reproduce themselves. Plants, too, must provide these complex compounds if they are to carry out the same functions. Getting proteins is a struggle, not only for animals and plants, but even for the microbes in the soil. Fortunately, for them and the plants, however, they can create or synthesize their proteins from the simple elements offered by the soil and the weather, if these offerings are complete. Man and his animals cannot create their proteins. They can only collect them according as the forms of life lower in the biotic pyramid, namely, the plants and the microbes, supply them. Accordingly, when the soils are naturally highly weathered, or have been heavily cropped for years, their supplies of the elements of fertility are depleted. On such soils the microbes and the plants are competing with each other for the fertility needed in creating proteins. Nitrogen—the symbol of the proteins—is a common deficiency prohibiting the legume crops from getting help to take nitrogen from the air through microbial activities on their roots. Even in the more fertile soils under cultivation, the many different microbial fires have already burned out much of the soil organic matter within which the nitrogen is held.

The open soils of the mid-continent with their excellent granular condition and high fertility to great depths let their organic matter burn out much faster than occurs in the more acid, heavier soils, farther east. Organic nitrogen is rapidly converted into the nitrate form in such deep, mineral-rich soils even when the moisture is low. This conversion moves the soluble nitrogen down ahead of the growing wheat roots. They catch up with this generous soil supply only late in the wheat's growing period by which the "horny" starch or high-protein grain is produced. This speedy conversion moved nitrates into the shallow well waters to threaten blue death to formula-fed babies by nitrate poisoning. It is also threatening extinction of high-protein wheat, and of good grazing for growing animals, now that big yields of soft wheat are hauled off the farm and even exported out of the country.

It was this high mobility of the soil nitrogen that has moved it into the grains and the livestock but out of the mid-continent all too fast. Unfortunately, the agriculture established now on those soils cannot keep on moving farther west. Crops must be grown there regularly but are now needing nitrogen supplements in their diets from the soil to the same degree as livestock is calling for protein supplements to their rations.

It is difficult to realize that the supplies of fertility in the mid-continent are declining seriously. But such is indicated clearly by the fact that beef cattle are now being fattened right in the West, and feeder cattle stay right at home to become fat enough for marketable beef processed right in the beef markets that have moved west too. It is also shown by the lowered concentration of protein in the wheat grain and in the green grasses being grazed by the livestock. When the "hard" wheat changes to a "soft" one, this means a grain that contains less protein and more starch. The less commonly recognized value of it as a "grow" food has been displaced by its more commonly emphasized value as calories or as a "go" food with fattening effects. With less nitrogen and other fertility elements from the soil, the crops may still make their carbohydrates to pile up the bulk and bushels because they cannot burn these for conversion into proteins of less bulk and better nutrition. Such soils put out the call for more supplements both as purchased minerals or chemicals for the crops and as purchased proteins for the livestock. They are no longer complete in their powers to create the grasses which truly create the cattle. Instead, they represent the problem of nutritional supplements for the microbes, the plants, the animals, and even man.

It is an almost forgotten adage which told us that, "To be well-fed, is to be healthy." It didn't tell us that to be well-filled with anything called food is to be healthy. We have too long neglected the declining fertility of the soil by accepting inferior substitute crops for the better ones that failed because the soil wasn't able to nourish them well enough to synthesize what they needed to survive and to make them quality feeds. In place of building up the soil to support the faithful crops, we have searched the world to find others. Those others were considered acceptable merely because they could make vegetative bulk where the predecessor failed in this.

But our cows have refused to accept these immigrant crops under that same criterion, because the cows couldn't produce calf crops on just any vegetative bulk being maintained at good tonnage yields per acre while fertility and nutritional values in those tonnages were gradually going downward. Now that beef can't be made so plentifully merely by turning the cows out to rustle for themselves, we look to rationing of protein supplements until there are scarcely any left on the market to be rationed. Only slowly will we realize that supplementing the feed for beef production is a call for our supplementing the fertility of the soil so that it will grow the grass crops delivering mineral supplements and protein supplements right in the pasture forages the cattle eat.

The protein in our own rations in the form of beef to supplement the carbohydrates in our potatoes and bread are not a problem premised on the need to roll back meat prices. Nor are the protein supplements to corn in our animal feeding a problem of rationing cottonseed meal, soybean meal, and other concentrates. Instead they are problems premised on the nitrogen deficiencies in our soils growing the plants within which the proteins are created from the elements and passed to us by way of the meat animals. Now that we have mined our soils of most of their nitrogen and other fertility in the virgin organic matter, we are gradually looking about for supplements and passed to us by way of the meat animals. Now that we have mined our soils of most of their nitrogen and other fertility in the virgin organic matter, we are gradually looking about for supplements from somewhere else. In the minds of some folks, this soil organic matter by which our past American prosperity was built is only a myth. In the experience of wheat for our staff of life, it is a shift from "hard" to "soft" wheat, or from a strong staff to a weak, if not a broken one. In the experience of our cattle herds it is a shift from plenty of beef to a scarcity of it, and from ready reproduction to difficulties in maintaining the herds even with many kinds of supplements and supports.

All of these troubles are gradually compelling us to admit that while we may manage much that comes after the creative part of agriculture, most of its truly creative phases still depend on what is in the handful of dust from which the assembly line of agricultural production takes off.

HOW SMART IS A COW?

How smart is a cow? Can we rightfully call her a dumb brute? As "lawyer for the defense" for the cow, we must remind you that she is smart enough to teach us she has better judgment of what is good eating for her than we have.

The cow refuses to eat broom sedge and tickle grass, which grows on soils so infertile nothing of more nutritional value can grow on them.

The cow also is smart enough to judge land by the feeding quality of the crop the soil grows, and not by the yield in bushels or tons. We're not that wise, yet.

The cow also is smart enough to pass her milk on to her calf without cooking or pasteurizing it, and without letting a lot of air attack it. We haven't become wise enough to learn what our handling of milk does to its nutritional values. If the cow can teach us something about food quality in relation to the soil that grows the food, we may well become close students of her behavior and put some questions to her.

At the Missouri Experiment Station we have been asking various animals about their feeds—particularly cows—and whether they can detect differences in the nutritional composition of forages grown on soils with varying soil treatments. They have been able to spot differences in hay in different stacks regularly for as long as eight years. After the last fertilizer application was made we can say that a cow is a good judge of the effects of fertilizers on the forage she eats. We can also observe that the effects of fertilizer may last as long as eight years.

These differences recognized by the cows have not yet been detected by the chemist. There is a suggestion, however, that the cow recognizes the better proteins produced in her feed as a result of the soil treatments. She doesn't show choice of feed according to the greater tonnage to the acre. Because she apparently is balancing the protein in amount and in quantity in her diet by her selections, we have raised some questions as to whether we will choose her carbohydrates according to the chemical differences between them.

Soil treatment can stimulate legume plants to make carbohydrates with sugars in prominence over the starch. Other soil treatments may make the starches prominent over the sugars. It is suggested by some tests that the cow prefers the plant with the sugars, or in the ratio of the sugar to the starches, hemicellulose and other less soluble carbohydrates.

By increasing the mineral element potassium, it was possible to produce plants with a high content of sugar, low content of protein. But when other mineral nutrients in addition to the potassium were increased in the soil, then the protein content was raised, the sugar content was lowered. Is the higher sugar concentration in legume forages an indication of low soil fertility? Is the high sugar content a temptation to the cow when she rushes into a field of legumes and gorges herself to the point of founder or other equally serious trouble? These questions suggest themselves.

When chopped wheat straw is refused by cattle, but is eaten readily when mixed with blackstrap molasses, there is a suggestion that cattle have a sweet tooth. In Oklahoma when sweetening was added to forages in the field, cows preferred the sweet crop. Shall we say the cow can be enticed by sweets to upset her digestive system as badly as the chocolate-gorging person will do?

Perhaps the cow is not so dumb! Perhaps she is telling us that her behavior is an act of desperation. Perhaps she expects these high-sugar crops to be legumes that also have some protein. Then because we have not built up the fertility to help make proteins in the plant as well as the carbohydrates, the cow may find she has been deceived. Shall we find fault with the cow if she appears to have been led astray by what appears to be a "sweet tooth!" In our limited knowledge of cows, shall we not credit her with judgment even in this case and hope for more knowledge of our failure in treating the soil to keep her from what may be an act of desperation which we provoked?

BEEF YIELD MEASURED BY SOIL FERTILITY

An average of 265 pounds of beef per acre in one season of grazing on the wheat-lespedeza plots at the Soil Conservation Experiment Station at Mc-Credie, Missouri, deserves attention with reference to the soil fertility involved as well as the enticing economy it demonstrates.

Cutting the labor cost to less than one-eighth, bringing erosion losses down to one-third or less, and yet producing more than common in money value per acre while selling less of soil fertility off the farm are highly desirable at any time, to say nothing of wartime. It is such facts that make the first and quick appeal in the soil-conserving pasture system of farming that grazes both the nurse crop and the legume following.

That some facts about soil fertility are at the foundation of such particular efficiency of forage as feed is a point that is likely to go unappreciated. Soil fertility measures the protein production per acre by the plants. Since meat is mainly protein, soil fertility likewise measures the beef yield per acre by the cattle.

The production of 265 pounds of beef per acre occurred in the wheat-lespedeza rotation, where 100 pounds of a 0-20-10 fertilizer have been applied annually with the wheat since 1938, and both the wheat and associated lespedeza crop have been grazed off. Then also, this fertilizer was going on land where limestone had been spread at the rate of three tons per acre at the outset of the trials in 1938 for all of these grazing plots using different crop combinations. Limestone is recognized as an essential treatment for good animal gains on this soil. It is also a requisite in making the phosphate and potassium go into the crop whether these plant constituents are considered in terms of their concentration of their total per acre in the forage. The applications of the fertilizer have been made annually for four times in order to get this amount of beef per acre.

In the case of two other crop combinations using timothy, including timothy-lespedeza in one case and timothy-sweet clover-lespedeza in the other as similar pasture systems, the beef production per acre was less. These two combinations were the same in their separate beef outputs per acre. Their figure was 225 pounds, which is less than forty pounds or about 15% below that on the wheat-lespedeza rotation.

One is prone to point to the differences in the cropping systems as the cause of these differences in beef production. But there, as in many other cases, the different soil treatments are the hidden and unrecognized causes. It is the lesser nutrient applications to the soil that must be considered as responsible for less beef where timothy was involved. These timothy plots have had the fertilizer applied only once and, though the application was 200 pounds an acre, it was put on in 1938. This was two years in advance of the first of the three seasons, 1940 to 1942, which gave the 225 pounds of beef per acre as the average in the three trials.

Here in the soil fertility applications lies the difference, namely 100 pounds of fertilizer annually on the wheat-lespedeza combination from 1938 to 1942 for a total of 400 pounds, but only 200 pounds once back in 1938 on the plots in timothy, and none through the years when the grazing trials were made. With the universal calcium shortage that this soil demonstrates by tests for its limes, and with the general crop response from applications of phosphate and potassium on it, these trials show themselves as a case where the animals are merely reflecting some results of soil treatments that are not appreciated when the focus of attention is on better beef yields, and the particular crop rotations only.

When a fertilizer mixture of superphosphate and potassium goes on regularly with the wheat seeding for pasture purposes, this treatment supplies many other nutrients besides these two found on the label. In fact, it is well to remember that superphosphate carries more calcium than phosphorus. In addition it carries sulfur in the equivalent of the phosphorus. These play their part in protein production in the animals as well as in plants. In addition to these gross elements there are usually some minor elements present that are coming in recently for attention for plants and animals. Manganese and zinc are commonly found in superphosphate to say nothing of copper and cobalt. It is this provision not only of phosphorus and potassium but of much more in terms of soil fertility that must be considered as exercising its effects on the plants in their synthesis of food compounds and therefore correspondingly on the beef production.

These different pasture systems deserve a prominent place in our farming for soil conservation value, but especially now for their labor saving value. The timothy combinations with legumes were better than the wheat-lespedeza for erosion control. They require labor attention only at the outset. They will hold from two to four years without reseeding and without any further labor costs. They stand up in grazing during wet spring seasons without serious damage to the future of crops which is not true in the case of grazing the wheat. The season of 1943 has demonstrated this feature very prominently.

These observations raise the question whether these timothy-legume combinations given additional and annual fertilizer applications might not offer higher beef productions as well as their other advantages. Perhaps with a few wet spring seasons thrown into this test, the low beef yield resulting from the trampled wheat will make the various crop combinations of more nearly equal value as beef producers even without the extra fertilizer on the timothy.

Soil fertility is involved in the business of animal production. In all of these plots lime had been applied at the outset when the systems were established. The soil treatments were different only in the more generous use of the fertilizer on the lespedeza-wheat combination associated with the higher beef yield. It is this difference that should be given casual importance here where attention is likely to be directed to the crop combinations and the crop sequences. This field demonstration supports other experimental studies pointing out most forcefully that the fertility of the soil measures the beef production per acre.

THE COW AHEAD OF THE PLOW

The art of agriculture, that is, agriculture in practice, is old. The science of it, namely, the understanding of the principles underlying the practice, is relatively new. The art has come to us slowly through the ages under guidance of the quiet but severe forces of evolution. Each long-lived practice is a case of survival by the strength of its recognized service. The science of agriculture has not been under test so long. It has often changed the art of agriculture. These changes were most pronounced where the natural resources, particularly the fertility of the soil, were ample to pay for, or cover, the costs of the mistakes connected with those changes.

Our westward march across the United States, to deeper, more fertile soils, has perhaps not impressed you as a case in question. We have not realized that bountiful soil fertility may have covered the costs of our errors in understanding the fundamental scientific facts of agriculture. Too much plowing, when we once recognized what the plow can do, and the resulting dust bowl, were only a temporary disturbance where the surface soil was so deep. A second dust bowl in the same place, however, might be a permanent disaster, if it should be the equal in severity of the first one.

The costs cannot be paid repeatedly by losses of this natural resource, namely, the soil. A science of agriculture given to direction from one center, and under national emergencies—apt to be overemphasized if not even propagandized—has left abandoned farms, and exhausted soils at the rate of one per family per generation in its make of westward travel. The art of agriculture has been more permanent, and less "progressive," if those are the costs of such distinction. The slower art of agriculture put the cow ahead of the plow. The science of agriculture, given to more speed, put the plow ahead of the cow. Cannot that science give more lasting profit to its agriculture by keeping the cow, our foster mother, ahead of us in vision as the reason for having any plow?

Primitive agriculture used the flocks and the herds, not the plow, as its symbol. Primitive agriculture was nomadic. It became a settled or a fixed one on those soils to which the cow had led the way. She served as the chemist assaying the soil fertility growing the forage that fed her and her owner. She had no fences to restrain her. She chose to graze on those soils fertile enough to make the satisfying feed of her contentment. She was not only ahead of the plow, she was ahead of the fence and other kinds of machinery that disregard her physiology and her nutrition in their designs according to the modern science of agriculture that is emphasizing economics, systems, politics, and all else except the nutrition of animals and man for their good health.

Some simple observations may prompt us to ask ourselves some questions and to stimulate our thinking about the soil as it provides all that the cow would need to have coming from it, and whether a soil needs to meet only the limited requirements of being just something to be plowed, or to be manipulated by other agricultural machinery. Forages grazed by the cow have been called "grass" and "hay." They have been something to be cut with a mower or chopped, baled, stored in the barn, and measured as tons of dry matter. More machinery to reduce the labor requirements of the farmer has been guiding the production of grasses and hays, when, in our humble opinion, that effort should be guided with the concern for more fertility in the soils for better nutrition of the cow by means of those feeds. The machinery has become primary in the agricultural picture. The cow has become secondary, if one can judge by the growing tractor population in contrast to the cow population.

When the plants we call "weeds" grow tall in the pasture while others are grazed closely and seem to be growing shorter, this is regularly considered a call for the mowing machine to fight the weeds. Instead when weeds "take" the pasture, that ought to be viewed as a case in which the cow is giving a new definition for the word "weeds." Careful observations of her behavior should raise the question in our minds whether she would suggest hormone sprays for the fight on weeds on a national scale. She is apparently telling us that weeds are not so much a particular plant species of bad repute within the vegetable kingdom. Rather, weeds are any plants making too little of nutritional values to tempt her to eat them for it. She lets them grow taller and tolerates the degree of her own starvation required to do so.

Recently a case came to our attention in which a herd of beef cattle was regularly going through knee-deep bluegrass and white clover on a virgin prairie, never fertilized and never plowed, to graze out the formerly well-fertilized, abandoned cornfield of cockle burs, briars, nettles, and a host of plant species considered our worst weeds. Most of them were the kind that are under legislative bans against distribution of their seeds. Here the cows were contradicting our plant classifications. They were disregarding what we offered as supposedly good grazing in the form of the bluegrass, and were going the greater distance to consume the plants we have always called weeds and even noxious ones. Such was their choice, though only when the weeds were growing on more fertile soil. These cows would scarcely recommend the use of hormone sprays to kill certain plant species we classify as weeds. They would recommend more fertile soils instead. Then, apparently, no plant would be called "weeds" by them.

When the cow breaks through the fence, is it her objective merely to get on the other side? A careful consideration of such cow behavior points out that she is not going from one of our fields to another one, both of which have had the fertility of the soil exhausted to a low level. Instead, she is going from one of those areas of our neglected soil fertility to the railroad right-of-way, or to the public highway. She is going to where the soils are still near the virgin, fertile condition. Those soils have not been mined of their nutrient stores. When as agricultural leaders, possible agricultural scientists, we plea for better fences to save valuable meat or milk animals from violent traffic deaths, is this not a failure to see the cow ahead of the plow? Is it not a case of putting the plow too far ahead of the cow?

On the coastal plains soils of the South, the automobile tourist is constantly confronted with the hazard of colliding with cattle crossing the highway pavement. They are not casual inhabitants there. They must be regular highway grazers. There are permanent highway signs to give warning of "cattle at large." These animals come out of the Piney Woods seemingly for miles on either side to graze this "chosen" strip of forage.

In spite of the punishable offense of killing one of them, many cattle in the South are accidently destroyed annually by the traffic. The high death toll results because the cows insist on grazing, not at some distance from the pavement, but right along its very edge on the grassy margin no wider than about one foot from the traffic-way. There the calcium, possible other plant nutrient elements, in the concrete mixture, diffusing through the adjacent soil or being taken by root contact of no more distant plants, apparently contributes a quality of feed the cows recognize and relish beyond that on any other part of the highway shoulder or the surrounding territory in the woods.

With the cow grazing so close to the pavement's edge and crossing so often to the other edge or side, she certainly is a serious hazard to the motorist. But she is a much larger hazard to herself. The larger number of fatalities to the cows as one of the two parties involved testifies accordingly. Here the mechanics of our well-developed system of transportation run not only ahead of, but counter to and in conflict with, the ancient agricultural art of letting this beast go out to select her own grazing under her judgment of its nutritional values reflecting the fertility of the soil growing it. While the machinery (the automobile more than the plow, in this case) is going ahead, the cow is not necessarily following it. She is being exterminated more often than the motorist, uninformed as he is of the forces responsible for bringing the dumb beasts as well as himself into this death-dealing situation.

We have been prone to ridicule the simpler arts of agriculture in the older countries, and the older civilizations where the plow and other modern agricultural machinery followed rather than preceded the cow. Just now we are engaged—on an almost international scale—in educational activities, savoring of a missionary nature, and aimed to bring these ancient agricultures up-to-date, at least in agricultural mechanics for mining their soil fertility. We are unmindful of the fact that in these older countries the agriculture was always highly pastoral. The arable agriculture never dominated so highly as we know it here, if the European manure pile in the front yard or the tank wagon flowing its liquid manure on the pastures and meadows dare to be considered as reliable indicators.

For us in the United States the plow has always been ahead of the cow. The plow has been agriculture's emblem. Arable agriculture and not pastoral has regularly been dominant. This was not so unexpectable in the age of farm machinery development, of more internal combustion engines, and of labor-relieving devices. It was the most expectable on soils containing great stores of reserve fertility. Our soils were of most extensive areas, very level topography, silty texture coming with windblown origins, high fertility in terms of its exchangeable forms on the clay, and rich in ready reserves of nutrients in the silt minerals brought as ample varieties from the arid West. Such soils naturally invited the plow and all kinds of machinery. Soil conditions of this type are natural temptations to convert them into cash crops, even for city suitcase farmers who would gladly escape the routine and daily work of milking cows and hauling manure.

Now that (a) the seriousness of erosion is being recognized; (b) the areas of fertile soils to be so easily exploited are gone; (c) the fertility decline is becoming apparent after being hidden so long under crop juggling; (d) the problem of protein supplements as animal feed and many of the troubles in animal production are being traced back to the soil and not alone to the feed store and the veterinarian; and (e) we are saddled with the responsibility of being Santa Claus for a much more inflated and hungrier world; we are coming to talk about less plow, and more cow as means to save the soil and to give us more meat and more milk.

While all these problems are too readily attributed to possible irregularities in economic and social arrangements, we are reluctantly coming to see the fertility of the soil underneath the whole picture. It was through the plow that we led the cow to soils contrary to her choice of the fertility there. The plow held her there just as the fence confines her to the deficient fertility in the pastures which are growing weeds in place of feed. In similar manner, our technologies of engineering have extended agriculture in its many forms of so-called "crop specialization" that are in reality cropping limitations because of limited soil fertility. Cotton farming is a case, sugarcane farming is another, forestry another, all of which are special kinds of farming that occur on soils of which the fertility would not entice the cow, and of which her assay would declare them too deficient to support her with good nourishment.

It was the plow ahead of the cow that took both of them to the once-forested soils in our eastern United States. It certainly would not have been the cow ahead of the plow. Would she select a forest site, clear it, and expect the crops to be good nutrition, when originally the Creator Himself could grow only wood there, and that only by the return annually to the soil of all the fertility in the leaves? We have allowed the mechanics of growing grass and feeding the cow to dominate our thinking so completely that the physiology of the plants and the physiology of the cow eating them have had little consideration. By way of the soil fertility we must be reminded that the cow is more than a mowing machine or a hay baler. Her physiology, and not just those mechanics are connected with, and dependent on, the fertility of the soil. She is not calling for merely tons of feed, and acres of grazing. She is calling for complete nutrition to undergird the reproduction of herself and for the establishment of the subsequent milk flow of high nutritional values for her calf. She is not aiming at establishing records of gallons of milky liquid and pounds of butterfat.

These visions and appreciation of the plants as physiological processes creating good nutrition for us and our animals do not come to us as readily as do our concepts of crop yields in terms of bushels of grain, and tons, or bales, of hay. The crop yields per acre as criteria of how well we are farming are the mechanical phase of farming, the plow part. The hays as good feed, rich in the vitamins, the minerals, and the proteins to grow healthy cows giving much milk are the physiological phase of agricultural creation, or the cow part. As these mechanical aspects became more and more prominent in our thinking, the physiological aspects were of less concern. By that token, the soil fertility as the foundation of agricultural creation has been disregarded and neglected. Under the prevailing agricultural criterion of more bushels and more tons, we have taken to searching for new crops, whenever a tried one began to fail. Instead, we should have been building up the soil fertility to nourish the failing one. The cow has never judged crop values according to crop pedigree. She has been telling us by her choices of the same plant in different places that the crop pedigree does not determine its chemical composition or its nutritional values for her. She does not follow the textbooks on feeds and feeding, accepting average values of chemical composition and digestibility per plant species with no mention of the wide variation in these respects within the same species. She more than the textbook is reminding us of variations in composition and digestibility per plant species with no mention of the wide variation in these respects within the same species. She more than the textbook is reminding us that variations in composition amount to as much as 1,000 or 2,000%, according as the crop is grown on soils of differing fertility. Surely the cow that is eating these variable samples of the same crop isn't taking them all at the same nutritional value. She hasn't ever heard of the mathematical mean or the average. She is given to marking out the differences and exercising her choice according to those variations. Hers is not the acceptance of the lot in terms of the average. With the cow going ahead, we too see the variations in chemical composition. But we see only the average figure when the plow is going.

That the soil fertility makes a tremendous difference in the chemical composition of a single grass species, and one considered high in the scale of nutritious grazing, was shown recently by the research of the Soil Conservation Service of the USDA, in their analysis of little bluestem of the western Gulf Region. Samples of this choice feed of the once-prevalent American bison were collected in close proximity. They showed a range in protein from 1.5 to 16%, in phosphorus from .03 to .31%, in calcium from .07 to 1.58%, and in potassium from .10 to 2.17% of the dry matter. The higher values were as much as ten times the lower ones in the cases of protein and phosphorus, 20 times for the calcium, and 21 times in the case of the potassium.

These were differences after the organic combinations influenced by, or containing, them had been destroyed by the ignition of the sample to leave only the ash. It says nothing about how widely the samples varied on the list of their organic compounds, like carbohydrates, proteins, and specific amino acids connected with the creative services these ash elements rendered. It is significant that these widest variations occurred in calcium and potassium, which are only recently coming into consideration because of their deficiencies in the soil. Calcium has long been ammunition in the fight on soil acidity when during all that time it should have been at the head of the list of needed nutrient elements on most humid soils for the nourishment of plants, animals, and man.

One of the major problems of the dairyman is that of providing all during the season, plenty of pasture in which the young grass growing vigorously is regularly available. For this, various successions of different crops on different pastures have been combined into so-called "systems" by which the cow is put on each pasture at the particular time when the young plants are growing rapidly. For example, one such system pastures bluegrass from April to June, Korean lespedeza from June to September, rye or winter barley from October to December, and then the dry bluegrass supplemented by cornstalks and hay during the rest of the year.

Such a "system" apparently disregards all that can be done by treating the soil as a help in feeding the crops to keep them growing. Bluegrass on fertile, well-watered soil permits regular cutting back to give new growth continually. But this regular recovery of growth after grazing, or cutting it back, calls for a continued and generous delivery of soil fertility. Timothy, on Sanborn Field, is given fertility has been growing during the entire season from almost the last snow of late winter through the summer and until the first snow of early winter. Regrowth by any crop cut back makes a demand on the soil fertility not commonly appreciated.

All of these facts ought to remind us that when our first pastures were established in this country, the soils were fertile enough to keep the crops growing under continual grazing throughout the year. The physiology of the plant calling for soil fertility is forgotten, as we mechanically juggle the cow from one pasture to another in such "pasture systems." Soil fertility must come in to lengthen the grazing season of any crop not only for more grazing of bulk, but also for the delivery of more nutrition in it.

While the idea of balanced ration for the dairy cow in terms of proper amounts of carbohydrates, proteins, vitamins, minerals, etc., has been commonly accepted, too few dairymen are yet talking about fertilizing their pasture areas to provide a balanced nourishment for the grasses according to the feed values they are expected to put out. The dairyman feeds the cow according to the milk she puts out. If he feeds his growing grasses he may well visualize the fertilizer as a balanced plant diet in the same way as he considers the feeds for the cow a balanced ration.

Unfortunately for our crops, the seeding of many of them has literally been a case of turning them out to rustle for themselves. Our soils have become so low in fertility that many fields are merely an empty feedlot, to which rain-water, air, and sunshine are delivered by the weather. Consequently, these meteoric contributions of carbon, hydrogen, and oxygen help the crops to produce carbohydrate bulk, but little more. Even legume crops are turned out similarly in the false hope, or poorly founded belief, that since they are credited with the use of gaseous nitrogen, the nodular microbes on their roots will help them make proteins from the same atmospheric source. In all of this thinking, the fertility required from the soil is completely forgotten.

Under such disregard of the soil factor controlling the feed grown for our animals, the problem of providing proteins has become a severe one. Plants make their carbohydrates first. These construct the plant as a kind of factory consisting of roots searching through the soil, and of branches holding out the leaves within which carbohydrates are made from air and water under sunshine power. Every plant must be making carbohydrates if it is growing. It is also making some protein at the same time. But unless the soil fertility is generously available as help for the plant's biochemical conversion of some extra carbohydrates into proteins, there will not be much of these coming along with the carbohydrates to make good feed for the cow. Only as the complete fertility outlay is in the soil for protein production by the plant, will that plant provide sufficient protein to balance its carbohydrates as a good ration for the cow. It is the shortage of soil fertility that brings the need for purchased protein supplements.

When the farmer has been telling us that legumes are "hard to grow," he has merely been confessing the fact that if the legume crops are to synthesize proteins while growing in the eastern half of the United States, some extra nourishment for them must be put into the soil in the form of lime and other fertilizers. Under the propaganda of fighting soil acidity, he was told that lime helped the legume crops because the carbonate of lime neutralized the soil acids. He should have been told that lime supplies the legume with calcium, a nutrient element needed for protein-building by plants even if no calcium is contained in the protein molecule.

It is on the highly weathered, fertility-depleted, and thereby acid, soils of the United States that the dairy cow is expected to provide us our daily needs of calcium in a quart of milk. Along with that calcium, there comes the protein portion of the milk, This fact points out that there is a close association between this inorganic element, contributed by the soil, and the big organic protein molecule assembled by the cow according as the plants have synthesized from the elements the constituent amino acids of it. Protein assemblage by the cow is merely the reflection of the synthesis of the amino acids by the plants according to the many elements of fertility in the soil. The weather may make carbohydrates, but the soil fertility determines the amounts of proteins, and their quality according to their completeness for nutrition in terms of required amino acids.

When the seed of this major grass, i.e. corn, is failing in its delivery of protein within itself, shall we not expect the corresponding failure in protein delivery in the other grasses grown in the same cornbelt and harvested at near maturity as hay? Is it possible that we are moving toward a pasture system of livestock farming because only the young grass is concentrated enough or complete enough in the proteins and all the nutrient substances associated with them to nourish our animals, and keep them reproducing? Then, too, are we not compelled to depend more on growing our own proteins because the once more common protein supplements are required for feed nearer to the points of their origin?

All these questions should bring us to connect proteins more closely with the soil under the animal rather than with only the animal itself. The cow cannot deliver proteins except as they are provided for her in the feed, save for the supplementary synthetic helps she can get from the microbial flora in her intestinal tract. The corn plant as a producer of the more complete array of the amino acids essential for the white rat and thereby presumably for the cow, suggests its capacity for delivery of such quality of nutrition limited to the germ of the corn grain. One needs only to feed the whole corn grain to Guinea pigs or rats to see how they eat out the germ first and no more, if the grain is plentifully supplied. The complete grain is deficient in the amino acids, tryptophane, methionine and even lysine. It is for the provision of these few deficient amino acids, then, that so-called protein supplements have always been, and must still be, supplied where the soil keeps plants from producing them.

It is significant to note that it was in connection with the increased concentration of these commonly deficient amino acids that soil treatments with the trace elements have been influential on alfalfa. Yet the trace elements are not component parts of the molecules of these protein constituents. Tryptophane contains the chemical structure known as the indole ring. This ring is broken down by the digestive activities of neither the human body, nor of the microbe, according to most late information. Can this fact possibly be connected in some way with the more common deficiency in forages and feeds of tryptophane? Methionine, another commonly deficient part of the proteins militating against their completeness, contains sulfur. This is an element not emphasized in fertility treatments of soil, though we have been unwittingly adding it through superphosphate, ammonium sulfate, and other sulfur-carrying fertilizers to say nothing of burning sulfur-rich coal. Sulfur was recently demonstrated as a beneficial fertility addition to some of our Missouri soils for bigger output of vegetation. It was shown beneficial, also in terms of more methionine in the crop growth. Shall we not give at least theoretical consideration then to the trace elements, not yet as known constituents of specific compounds in the body structure but possibly as tools in the fabrication of essential compounds by processes still unknown and still not recognized? As tools, then, we cannot expect to find their amounts directly correlated with the magnitude of the products they affect, any more than one would expect the number of milk pails, to be an index of the number of cows being milked. Counted milk pails of a dairy are no measure of how much milk it produces per day. Only by means of more refined research, and much of it through the animals used more directly to assay the services by the soil, can the nutritional services of the trace elements be elucidated. In this the cow must go ahead of the research machinery of which even she is seemingly a more refined detector, including some of the finest of research tools. Even the spectrograph.

We are slow to appreciate the cow as the symbol of the physiological requirements of all life and can be properly nourished by agriculture only when the soil is properly plowed, and is treated for supplying via the plant all the essential elements and compounds that are food for growth and reproduction. Modern agriculture is threatening to put the plow so far ahead of the cow that crops will be grown for little more than their bulk of vegetative delivery. Even that is more and more under neglected soil fertility so that she can scarcely find enough nutritional values in her feed to survive. Machinery of many kinds seems to be conniving with the plow to have us forget the cow entirely. In place of drinking milk we seem to persist in the unweaned habit on a national and even international scale, of sucking continually on the carbonated drink bottle. Fats from all sources are replacing butter by the help of machines that make hard fats soft and soft fats hard, and give us chemicals for artificial colorings that rival the carotenes themselves. Cheese, that once spoke with masculine accents and strength of its own, is now being replaced by the many "cheese foods." These are too effeminate to be sandwiches with rye bread and accompanied by the customary drink once gulped down with it in quarters not commonly frequented by women. Some 20 years ago we took to plowing under prenatal pigs. That performance has apparently become such a habit that it has moved the plow to cover the cow too. While machinery is always a helpful tool under good mental guidance, still that contribution by the mind dare not be too small a part in that partnership of mind and machine. Surely we must provide some knowledge of the soil as nutrition if the cow is not to be deleted completely by the machinery that ought to follow behind rather than go too far ahead of her.

Unfortunately, as soon as research helps us gain a bit of physiological knowledge of nature's activities of agricultural significance, we make machines to capitalize on it. We become so engrossed with the running of those machines that we cease to search out more knowledge of more physiology. The young scientist who made the discovery no longer continues his research. Instead he becomes an administrator, a director, or a business manager in the commercial development of his discovery. He leaves the cow, but he takes to the plow.

The cow has served in this discussion as the symbol of all the life forms with the creation of which agriculture deals. She represents the science, the organized knowledge of that life. Only as we understand the physiology of the microbes, the plants, the animals, and ourselves, can we fit all of these life forms on the foundation of their nutrition and thereby of their creation, which is none other than the soil.

The plow has served as the symbol of the inclination to move to mass production, or to set agriculture going and let it run by itself under no more serious criterion than the collection of big monetary values by means of it. That philosophy of agriculture is threatening to be the Frankenstein monster about to kill agriculture itself when it starves out all the life forms on which agriculture depends. It is about to leave us hungry with only eroded, and barren soils because their fertility has been neither restored nor maintained.

The Experiment Station's research must be challenged by some of the fundamentals that are not measured completely by criteria including no more than yields as bushels or cash returns. Qualities that deal with life, not quantities of materials alone, must be emphasized. In that research the farmer too must share some of the thinking responsibilities. He must do more than just ask the Experiment Station his many questions, and expect a practicable answer merely for the cost of inquiry. The experiment station cannot, and dare not, in terms of democratic principles, think for the farmer. But any researcher with better agriculture as his objective should take delight in any opportunity to think with him. As more farmers think about the fundamental processes of creation by which all that we call agriculture is supported, our thinking will not be contented with the machinery, the costs, the prices, and the speculative aspects of agriculture alone. It will invest itself more in the understanding of production that depends on the fertility of the soil. It will not be disturbed by talk of surpluses, that are only surpluses in quantity. But it will be seriously disturbed by shortages in quality, especially of foods so closely associated with reproduction of life as milk, meat, and eggs are. Such thinking by the farmers cooperating with the research men of the Experiment Station will keep the cow ahead of the plow, with decided advantages for her as well as for us.

ANIMALS KNOW GOOD FOOD!

Our farm animals are telling us much—but they speak in a foreign language. The situation reminds one of the fellow who said, "Money talks but so far as I am concerned, it speaks a foreign language."

The case of a famous Missouri donkey (or jack) reveals the story of the soil as the basis of animal production. This animal came from that part of Missouri where donkeys were made famous. The great breeder that used to breed them unfortunately isn't with us any more. The farm on which that donkey breeder made himself famous in Missouri was known as Limestone Valley Farm.

We must not forget the donkey is native and wild in the regions of low rainfall, where there are high concentrations of calcium in the soil. He is an animal that is surefooted and accustomed to crawling from rock to rock. He is a most trusted saddle animal for you, if you want to go down into the Grand Canyon. The burro, or donkey, is given to a diet that is high in calcium and in those alkaline earth elements, because of the soils that grow his feed. Move him out of Missouri, or move him off the Limestone Valley Farm in south Missouri to northwestern Missouri where this one under discussion went as a colt—because he had a noble pedigree and sold for $5,000 on the strength of that pedigree— and another severe case of rickets comes up, as this one did.

When this donkey was moved into that part of Missouri which we say is great bluegrass land, his pedigree couldn't keep him free from this terrible affliction. The animal was deformed under that level of nutrition. In spite of all that could be done for him, a shift from one soil to another was more significant than breeders, veterinarians, and others dealing in animals ever surmise.

Observations of choices by animals at large, and under control, using feeds grown on soils given different treatments, point out the animal instincts for selecting not mainly carbohydrates for laying on of fat, but proteins, inorganics, etc., for body growth and reproduction.

Animal behaviors are now undergoing much closer observations. Reports of animal choices according to the soil are becoming numerous, and the various explanations of the casual relations are legion. As assayers, our animals are making excellent biochemical contributions. The animal's instinctive choices of its grazing according to the soil fertility, but its ingestion of almost any plant on fertile soil—irrespective of our classification of it as a weed or as a commonly accepted herbage—points out the animal's acceptance of crops according to the soil's synthetic services through them and not to the reputation founded purely on plant pedigree or scientific name.

Rabbits fed the same variety of hay from the major soils with and without treatments, in an area as extensive as the State of Missouri, demonstrated wide differences in growth, bone development, and body physiology according to the soils and their treatments. Originally as uniform as possible in breeding and selection, they were widely different for the various soils after a short feeding period. We are slow to see the soil in control of the characters of the animals, while we cling to a traditional faith in breeds and pedigrees as determiners of animal body form.

Deficiencies in nutrition are gradually coming into consideration as causes of poor health. However, it is at a slow rate at which those deficiencies are being considered as originating in the soil; either directly as a shortage of indispensable inorganic elements, or indirectly as compounds fabricated through their help in microbes, plants, and animals.

In order to test the hypothesis that such a baffling ailment like brucellosis in animals and man may be due to deficiencies, particularly of the trace elements in the soil, Ira Allison, M.D., Springfield, Missouri, volunteered to cooperate and try some trace element therapy on afflicted humans as was suggested by Francis M. Pottenger, Jr., M.D., of Monrovia, California. Dosages of manganese, copper, cobalt and zinc under enteric coating have been used on more than 1,000 brucellosis patients to date. Other ailments have also been included. The changes in the blood picture and other clinical aspects, coupled with the patients' reports of improvements in health, have been most encouraging with hope for relief from this disturbing ailment. This hope also includes amelioration of other ailments, like arthritis, allergies, anaemia, eczema, angina, etc., as possibilities.

We are slowly coming to believe that much that we now call "disease" should be more correctly labelled nutritional deficiency. As this is more widely granted, there is a growing appreciation of the importance of soil fertility in nutrition in general. Such recognition is making each of us more ready to take some responsibility in conserving the soil by which all of us want to be well fed and thereby healthy. It is also putting less emphasis on fighting disease and more emphasis on its prevention by better foods grown on better soils.

WILDLIFE LOOKS FOR BETTER NUTRITION

Because our wildlife looks after itself, we do not appreciate how it, too, struggles with problems of nourishing itself completely. Only as animals accomplish this do they live and multiply. Otherwise, they become extinct. They live, then, in the areas which are properly nourishing them.

In our domestic animal pattern we might at first imagine that animals are merely scattered according to our accidental distribution of them. Quite to the contrary, we have pork as a fat-producing animal in the eastern United States where carbohydrate crops, like starchy corn, are its major feed. This is on soil that was originally forested. In its virgin condition it was producing mainly wood, or fuel, as its crop. Today it is producing mainly fuel foods and fuel crops. These are high in calorie values. They are low in protein; even as feed for hogs corn must be accompanied by some protein supplements.

On those soils the original wildlife had its struggles with problems of nutrition. Wildlife in the forests tells us of struggles for calcium and phosphorus. The porcupine in the woods of Minnesota was observed as it was eating the antlers of deer dropped there. Antlers disappear quickly in the forests on lime-poor soils because the other rodents, like mice, and squirrels as well as the porcupine, supplement their calcium deficient diets by calcium concentrates in the form of bones of preceding life forms. This is their struggle in order to guarantee complete nutrition.

In the same area today, cleared of most of its forests, the gray and red squirrels are commonly seen eating bones. Seen on the window ledge, they have been carefully observed eating old dry bones, so aged and so dry that obviously it was the bone itself that appealed to the squirrel. This suggests that the feeds produced in the neighborhood for these animals are an incomplete diet in terms of the lime and phosphate which only the soil can supply.

It is suggestive to note that it is during the season of pregnancy that most of these reported observations have been made. Perhaps during part of the season the struggle is not so marked; hence, the animal needs only supplement by such perverted behaviors of eating the remnants of its predecessors' bodies during a limited period of the year. Crops too low in lime and phosphate, and with only fattening power as feed for hogs, call for bones as supplements in case of their feeding wildlife; just as legume crops on those same soils call for lime and phosphates if they are to produce the protein-rich feed values and successful growth.

In the western United States where high-protein wheat has been a common crop and where grasses were once the virgin vegetation, antlers of deer accumulate. It was in that area where the buffalo roamed. With the summer he moved northward, and with the winter he moved southward. He did not move much to the east of his chosen ares, not even into the eastern limit of the big bluestem except possibly for Kentucky. In that state he was in limited numbers and limited area originally where the racehorses have made themselves famous and established the annual derby. He was also in some of the more productive valleys of Pennsylvania—to suggest that he too was struggling to find his complete nutrition. This was according to the pattern of the soil.

It has been reported that deer quickly take to burned-over forest areas on the first signs of growth recovery there. Can they recognize the ash as soluble fertilizers making the grasses and other growth more rich in the inorganic elements and thereby more proteinaceous? Can this burning get rid of the excessive woodiness in the decaying organic matter that keeps the inorganic essentials tied up in microbial life in competition with plant life? Does the deer get out into the burned-over clearing and risk its life in the struggle for complete nutrition? Deer as marauders in our fertilized fields and well-kept gardens are risking their life in their struggle to find what they need to survive—all their fear of the human enemy notwithstanding.

The wildlife in its struggle for complete nutrition has set up its population pattern as the result of—and thereby in agreement with—the soil fertility pattern. All this tells us that our wildlife is just another crop of the soil and only as we manage the fertility of the soil to feed the wildlife will we have this crop in abundance.

SOIL TREATMENT AND WOOL OUTPUT

The wool of the sheep, much like muscle, is a protein product of growth. While fat is associated with wool fiber in what is commonly called the "yolk," this fat is the secondary rather than the primary part in the growing of the fleece of wool.

In the production of wool, for too long a time, we have been concerned with the quantity the sheep produces. Only recently has the quality of the wool fiber come to attention. And much more recently, we have begun to consider the quality of the wool in relation to the fertility of the soil growing the forages on which the sheep are grazing. We have now begun to realize that the fertility of the soil may come in, to control the wool to degrees not generally appreciated.

In some trials with hays grown on soils with different treatments, the wool of the sheep scoured and carded out in good order, when the hay was grown on a soil treated with both lime and phosphate. It could not be carded and the fiber was broken, in the attempt at carding, when the hays fed the sheep were grown on the same soil type given only superphosphate. In addition, there was little yolk in the wool on the sheep in the latter and a generous amount of it in the former. Here the quality of the wool fiber of protein make-up, and the amount of fat associated with that fiber, both varied widely simply because the lime deficiency in the soil had or had not been corrected by an application of limestone.

More recently the regularity in the crimp, or the wave, in the wool fiber of the Merino sheep has been correlated with the presence or absence of some of the trace elements, especially copper and cobalt. In South Australia this was very noticeable in the wool from sheep grazing on the grass on some of the calcareous sands. The drenching of the sheep with from seven to ten milligrams of copper associated with the one milligram of cobalt daily restored to its uniformly wavy appearance the so-called "steely" or wire-like wool fiber.

The color of the black wool of black sheep is also related to the presence of these trace elements, especially with the copper. Australian black sheep, grazing on pastures on these coastal soils, lose the black color to become gray. When given copper and cobalt in the feed, or the drench, the black color is restored.

In many parts of the United States it is commonly said "Black sheep never stay black. They always turn gray." It might well be questioned whether this is not merely saying that our soils are copper deficient, and possibly even cobalt deficient. We have only recently begun to think of these so-called "trace" elements and their importance in the body's production—not of fats, but of its proteins, even of the different amino acids constituting the proteins.

Some studies on the essential amino acids making up the proteins in alfalfa, for example, suggest that the trace elements put in the soil are effective in pushing up the concentrations of these protein constituents created by the plants. Sulfur put on the soil suggests that this element is too low in supply in the soil to let the plants make all the sulfur-containing amino acids they would otherwise make. Magnesium as a soil deficiency also comes into prominence now in connection with possible failure of our forage crops to build the proteins complete, so far as the required amino acids are concerned.

Now that we are measuring proteins more commonly in terms of their nutritional completeness with reference to all the amino acids, rather than by calling all nitrogenous products proteins, the soil deficiencies for plants creating these acids are coming into the foreground.

We need to recall the fact that animals don't synthesize proteins from the elements, including those in the fertility of the soil. They only collect them from the amino acids created by the plants. Sheep and their protein output in the form of wool are telling us that the fertility of the soil, and our treatments of the soil, reflect themselves quickly in this product. All this tells us that the creation of what carries life itself, namely, the proteins, goes back to the soil.

COWS KNOW NUTRITION

Wild animals evidently know their own proper nutrition for good health. They are not reported as exterminated by epidemics, though their numbers may be reduced by disease. Domestic animals also tell us of their wisdom in nutrition when they break through the fence to go from the time-worn farm soil to virgin territory on the highway or the railroad right-of-way.

They will also leave the feed in one manger or rack for another, choosing the one of potentially higher nutritive value. They will select the better treatment or medicine from among several for an ailment, a dysfunction, or a deficiency, even to improving on the advice of the veterinarian, if we cooperate in their making of such decisions.

That such seemingly uncanny behavior is a fact has been reported by Mr. E.M. Poirot of Golden City, Missouri, a farmer-naturalist whose keen observations and logical thinking about animals make him emphasize agriculture as a biological performance first, and an economic one as a sequel.

In the dry season of 1955, when grass in his virgin prairie pasture was offering no verdure in early August, there was a serious and sudden outbreak of typical "pinkeye" (infectious ophthalmia). This affliction of watery, pink,, and seriously irritated eyeballs occurred in about half the herd of 300 Holsteins. It showed no discrimination as to the animal's condition, age, or sex, except that the suckling calves seemed to be immune. Under the drought, and the animal impatience with flies and heat, naturally the irritation from dust, the possible injury from crowding under shade, and the customarily suspected infections would be called to mind first in the diagnosis.

Such was the thought-path followed by the veterinarian who recommended the treatment of each eye with terramycin. But it was not the path of thought taken by Mr. Poirot. He recognized quickly all the hazards of handling so many animals in the roundup, in putting them through the chutes, and in trying and treating each with the antibiotic, especially in the heat of summer.

He reviewed his knowledge, accumulated in the past 35 years, of both wild and domestic animals and recalled the adage, "To be well fed is to be healthy." He set up, as a postulate for test, the converse. An affliction like pinkeye might well be a case of nutritional deficiency, especially of trace elements of the soil, even if only an indirect effect.

Approximately 35 years ago Mr. Poirot had undergone the struggle of dosing the animals and had borne the losses of eyes and animals in spite of professional help, which included vaccines made from cultures from the infection in the animals' eyes. Five years later he had carried out his own homespun research with lumpy jaw (Actinomycosis, reported to be caused by *Actinomyces bovis*) and discovered that it was prevented and cured by providing the cattle with compounds of the trace element iodine mixed in their salt. He had been using this preventive and cure ever since. He deems it necessary because his soils are seriously deficient in both the major and trace elements of fertility, which may cause important changes in the crops they produce.

About 20 years ago the use of copper sulfate for internal parasites provoked the question, "When internal parasites are not killed, but are driven out alive by such dosage, might not the real cause of their extended stay in the alimentary tract be connected with deficiency of copper as nutrition?" As a result, the feeding of copper sulfate, when there was no verdant feed, had been his regular practice until about ten years ago when the copper, at the rate of ten parts was combined with one part of cobalt with some added iron compound. All was combined with ordinary salt.

Long before the concern about the deficiency of trace elements in his soil, Mr. Poirot had discovered the lack of calcium, magnesium, and phosphorus in the grass and soil of his virgin prairie. His cattle selected one of four haystacks from which to feed. About one-sixth of the hay came from an area given calcium and phosphorus as soil treatment eight years before. Since observing the cattle's choice Mr. Poirot has kept calcium and phosphorus in salt form where the animals could get it, with steamed bone meal as a separate offering along with ordinary salt for its sodium and chlorine.

It was these past experiences which led him to set up the most recent exhibition of cows as diagnosticians when they demonstrated again this winter that pinkeye is not of infectious origin (though the eyes are infected), but has its origin—on his farm—in deficiencies of possibly one or all of the three elements, cobalt, copper and iron.

A severe outbreak in the dry summer of 1955 reminded him that he had not been offering the cattle this trace element mixture. Upon putting it out for them again the trouble was cleared up in a little more than two weeks. Then, after another period of no feeding of these elements the reappearance of the affliction suggested separation of the different trace and major elements a bit farther in the animals' discriminations.

Bone meal, monocalcium phosphate, triphosphate of magnesium, muriate of potash, ordinary salt, and a mixture of copper, cobalt, iron and salt represented the separate offerings. The cattle chose the cobalt-copper-iron- salt mixture and the bone meal, preferring these to the other forms of phosphates and of salt. The cows were correct in their diagnosis and prescription, since again the pinkeye disappeared with no loss of either eyes or animals.

Cattle can teach us, if we let them, that they can diagnose, prevent and cure their dysfunctions caused by the deficiencies which we so readily label "disease."

And very slowly do we seem to realize that nutrition is dependent, via the inorganic elements, on the soil that grows the organic combinations which we use as food and feed. By this implication our gardens should have first consideration as the home source of health because of nature's help in providing good nutrition.

BIGGER BONES DEMAND BETTER SOILS

Bones suggest skeleton as the stiff, mineral framework that supports the overlay of muscles and other soft body tissues. Bones have had little consideration as accurate recorders and responsive reflectors of the fertility level of the soils growing our foods. Bigger and better bones of farm animals have deeper meaning than just more mechanical strength of the body. They are a record of the animal's nutrition as this gives good or poor growth and reproduction, all according to the soil fertility under the crops supplying the feeds.

Bones consist mainly of compounds of the mineral elements, calcium and phosphorus. These two nutrient factors for all life are what we put into the soil when we apply lime and phosphate on the land. If the soil is deficient in these two elements, then poor crops and poor animals with weak bones, irregularities in reproduction, and other body troubles result. Unfortunately, most of our soils need these treatments for the health sake of all life. But what is more unfortunate is the fact that the evidences of it in the form of thin, light bones remain hidden and these soils and health deficiencies continue to be neglected.

The bones as evidence of such were brought out clearly in the case of two

Jersey heifers. Both were fed the same ration that was normal in calcium but low in phosphorus. But one heifer was given a supplement of mineral phosphorus. It was this animal of which the leg bone was twice as heavy, of thicker walls and of less porous construction near the joints.

Bones are not just dead parts, or just reinforcements, of the body. They are physiologically active. Production of red corpuscles occurs there. They are the reserve supplies of calcium and phosphorus from and to which these two essential elements of life move when supplies of them in the food are short or abundant, respectively. Shortages and abundance depend on the soil.

Bones cannot be built from fattening foods, or the "go" foods made from air and water. Building better bones calls for the mineral-rich, proteinaceous foods—the "grow" foods—compounded by the plants demanding better, or more fertile soils.

Better soils—better bone structure.

ANIMALS RECOGNIZE GOOD SOIL TREATMENT

Measuring the effects of lime and fertilizer additions to the soil only in terms of increased bushels or tons fails to give us the fullest evaluation of what such soil treatments are doing. If we will turn, however, to the animals that must eat the forage or the grain produced on the treated soils, we may discover that they point out by their tastes and choices some effects that we do not appreciate. Animals will recognize soil treatment effects too small to be recorded as weight differences, as well as improvements in the crop quality lasting long after the soil treatment may have been forgotten.

An illustration of this fine taste for improved soil fertility even in animals of supposedly depraved appetite is reported by Mr. Burk, the county extension agent of Johnson County, Missouri, who cites the experience of Cliff Long. Last fall Mr. Long turned his hogs out to hog down some corn. This was on a field in the far corner of which, 60 rods distant from the gate where the hogs entered, limestone had been used on a small area some years ago. To Mr. Long's surprise, the animals went back and forth from the barn lot, the water, and the tankage supply at the gate, through about 80 rods of corn on unlimed soil to consume first the corn on the limed land.

"I don't believe it," you may say, but had you seen the hogs as Mr. Long and his neighbors saw both hogs and the field, you would no longer express doubt. You would take the testimony of these hogs about soil fertility even if the chemistry of the effects of limestone on the soil and on the corn plants may seem like only so much "hocus pocus" to you.

As further testimony, the cows on the farm of E.M. Poirot, Golden City, Missouri, declare their ability as connoisseurs of finer fertility differences reflected in the spring barley pasture. Mr. Poirot's barley was drilled last fall with an application per acre of 100 pounds of 32% superphosphate through a tractor drill. This operation was carried out by drilling in "lands" and left the corners of the field rounded and partly undrilled during the main operation. In order to drill out these corners, a few rounds were made from the corners diagonally out into the field. This drilled a double dosage of the phosphate fertilizer over part of the field. In one of the drill rows in the main part of the field there was a very heavy application of fertilizer when a fertilizer gate remained open. This barley row was already much larger last fall.

This spring when the barley was pastured, these areas given the heavier application of 200 pounds of phosphate per acre were taken by the cattle first, and that part of the field given only 100 pounds was left ungrazed distinctly right to the drilling line. The one drill row in the main part of the field given extra heavy treatment by the open gate on the fertilizer drill was also singled out and taken first by the cattle.

Perhaps you may ascribe such discrimination to some other item than the fertilizer, but the cow knows her feed and can distinguish between difference in barley feed quality by soil treatments as small as hundred-pound applications of phosphate, as this report clearly demonstrates. Plants are sensitive with responses not recognized by the eye.

As a further illustration of the forage differences recognized by cattle, Mr. Poirot cites a case where the cattle selected a small area within 190 acres of virgin prairie pasture where 500 pounds of limestone per acre were drilled as a demonstration in 1928. He says, "The cattle have stayed on this smaller limed area this spring rather than graze over the entire 190 acres." Here in this virgin prairie, the addition of so small an amount as 500 pounds of limestone drilled but lightly into the surface has left an effect on the virgin, native forage, which even the supposedly dumb beasts still can recognize after almost 11 years. It suggests that liming some of our older pasture by an early spring drilling that doesn't tear up the sod may make differences appreciated by the cow even if we don't recognize them.

The animals confined to our farms are exercising the best of their judgment to make good use of their feed and are pointing out, whenever possible, that the feed is better as the soil fertility is higher. So far, they have pointed to lime and to phosphates as essential in making the feed better to their taste, and doubtless, to their more effective growth, milk production, and other body functions. As fast as we set conditions that will let them demonstrate that they can pass even better judgment than we can, they may point to other fertility items in which our soils are deficient for best feed production. Given a chance, our animals may teach us to appreciate our declining soil fertility.

BETTER SOILS MAKE BETTER HOGS

When Professor Evvard of Iowa once said, "If you will give the pig a chance, it will make a hog of itself in less time than you will," he was the first to point to the hog as a capable assayer of the nutritional values of the components of its ration. As a feed chemist the hog is demonstrating a much more refined technique than we might believe when just watching it as it balances its carbohydrates against proteins and minerals from the supplies of these we have put into the self-feeder. In fact, the hog, like many other animals both wild and domestic, is capable of selecting its food according as these reflect the higher fertility of the soil growing them. Have you ever thought that better soils make better hogs because of the higher nutritive value per unit in the feeds through higher fertility in the soils growing them?

"Yes," you may say, "the hog is closer to the soil than most other animals, since it wallows in the mud puddle in the warmer summer and usually roots up the sod during the spring. It eats so much of its feed right off the ground too, so of necessity it is ingesting much soil and gets much that is, purely mineral." However, the question may well be raised as to just why the hog roots; when it does so mainly in the spring after a winter on low-protein and mineral-deficient feeds; then too, when it doesn't root later in the season after it has gotten green feeds; and when it seldom roots if it has access to good green alfalfa. We are slow to believe that the hog—once a roamer but now confined by fences—is trying to report the poorer quality of feed because of the declining fertility of the land also enclosed inside of them. What we might be prone to consider just plain "cussedness" in the hogs breaking through the fence to the highway and their insistence on rooting may be our failure to appreciate (a) what all is demanded to give most efficient feeding of them, and (b) their behavior as indicators, or helps to tell us what in the way of extra nutrients they need.

The Hampshire, which in its domestication has not yet been turned so completely to "slothfulness and gluttony" as possibly some other breeds, is credited with being "a good rustler." Might this not be merely a good ability to search out the essentials for making up its own well balanced ration? Along with this ability as a rustler goes the reward for it, namely, the larger number of pigs per litter for which this breed has the claimed and granted reputation. All this suggests that if we will give the pig a chance to meet up with better soil fertility over which it feeds, it will not only make a hog of itself quickly, but will make more hogs as offspring as well.

That the hog will select its feed according to the fertility of the soil growing it was forcibly demonstrated in Missouri some years ago on the farm of Cliff Long in Johnson County. In attempting to "hog down" some corn, the tankage and water were kept at the gate of the 40 acre field. The hogs came out for water daily and went back and disappeared. The absence of indication of their activity caught Mr. Long's attention. He was becoming concerned at about the same time that his neighbor, across the road along the opposite side of the field, reported to him about the clean job of corn harvesting the hogs were doing.

On examination of the area hogged down Mr. Long found that the hogs were carefully working to the very border lines of the small part of the field where several years before he had limed and fertilized the soil to grow some alfalfa. The hogs had roamed the forty acre field; they had taken first the grain grown on it. They had mapped the soils of this field according to the differences in their fertility and with an accuracy no soil surveyor could even imitate.

"Just what is the characteristic of the corn that is being searched out and recognized by the hogs?" you are already asking. Much has been said about it being the minerals, because hogs eat soil, coal and other minerals, and too it was mineral fertilizers that were added as the original soil treatments for alfalfa. There could be nothing else but minerals though not necessarily directly, responsible for making this particular soil area in the forty acre field different from the rest of it. Of course, if the chemist should analyze the grain he would burn it down to ash for his determinations in search of differences. Consequently any difference would be reported as one of the minerals.

There is doubt, however, whether the hog is assaying its feed wholly according to deficiencies or differences in the ash or mineral constituents. The hog is taking part of its food as carbohydrates for energy more than these for their ash. It takes food for its protein content more than for the ash associated with this dietary constituent. Does not the plant do more for the hog than just haul minerals to the trough and the self-feeder? Certainly it is within the plants—and only by the processes there—that synthesis occurs of the amino acids out of which proteins are constructed by the animal body. It is of proteins that muscle is built. It is the plants that construct carbohydrates through photosynthesis. It is the carbohydrates that help the hog to lay on its fat and that give energy to its body. These services performed for the animal by carbohydrates and proteins result from them as complex compounds synthesized by the plant's life processes and can not be performed by the mere presence of the plant's ash-comprising minerals. The hog is not assaying for the mineral contents of its feeds. It is assaying for the degree of balance which its feed contains of the many complex nutritional substances—carbohydrates, proteins, fats, minerals, vitamins, etc.— that only the plant can put together and then most effectively only by the help of a good store in the soil of all the elements of mineral fertility.

It was Professor L.A. Weaver of the Missouri College of Agriculture who demonstrated the significance and good performance of high soil fertility in terms of pasture for hogs. He used a series of different crops as grazing for them and used corn as the supplement. If this series of crops is arranged as they require less fertility from the soil for their growth—or as "they can be grown more easily" as it is more often said—his results show that they also make less pork per acre. Alfalfa, which is considered the hardest to grow, is at the top of the list of the crops with its recognized high pork production per acre of 591 pounds. Then the other crops follow in this order, namely, red clover 449, rape 394, sorghum 275, bluegrass 274, soybeans 174, and cowpeas 149 pounds of pork per acre. These are their capacities to support the hog in its inclination to become pork when corn was a supplement. In respect to the amount of corn, as bushels, required to make a hundred weight of pork, these hog pasture crops listed themselves with the following figures: Alfalfa 5.5, red clover 5.2, rape 5.3, sorghum 7.1, bluegrass 7.3, soybeans 4.6, and cowpeas 5.2 bushels.

When we say a crop is "hard to grow" because it requires fertilizers like lime, phosphate, magnesium, potash, boron and other nutrient elements, we forget that it is just these "hard-to-grow" crops that are more nearly manufacturing all that it takes to convert a pig into a hog quickly. Alfalfa, for example grows naturally today on the soils that made buffaloes as big and numerous as the Forty-niners found them. No one ever reported the buffalo as having a separate source of protein supplements. He found his protein synthesized by the short grass that bears his name. But that grass was undergirded in its protein manufacture and its mineral delivery by an ample fertility supply in the soil that had not been leached out by heavy annual rainfall. It had not been weathered out to the degree where forests or wood was about all that the plant could synthesize.

The hog as producer of a higher percentage of its body as fat, than is the case with the buffalo, or cattle and sheep, is able to do better than those on soils that grow carbohydrates, like corn, more commonly and more easily than they grow protein, like hard wheat. Nevertheless, the hog must also have its proteins. Unfortunately, for the better upkeep of our soils, proteins are too commonly considered as something that must be purchased in a bag rather than grown right on the farm in our non-legume crops by giving the soil some lime, phosphates, and legumes or other nitrogenous fertilizers which increasingly greater areas of the soils now require to produce it.

That the hog discriminates between the corn or the wheat, for example, according as it was grown on soils with the different fertility levels or different fertilizer treatments has been demonstrated very positively. The hog has been recognized under experimental tests as a very good assayer of differences in the soil fertility growing its feeds. The same hybrid corn or the same variety of wheat was grown for example, on three plots of soil given (a) lime, phosphate and potash, (b) lime and phosphate, and (c) lime only. One hundred pounds of the grain harvested from each of these plots is put into separate compartments of the self-feeder and the hogs allowed to consume these according to their choice. Regularly the remnant amounts are weighed, and the compartments refilled with weighed quantities, but not in the same order or position of the different grains in the self-feeder. This is the procedure used to test other kinds of soil treatments. This method of asking the hog to assay the values of different fertility treatments does not report them as increases in bulk such as bushes or tons per acre. Rather, it is the pig's report on the nutritional quality of this part of its feed and the pig's suggestion on what we can do in helping it to make a hog of itself most efficiently.

What then are some of the reports by these pigs as endorsers of our efforts to improve the feed at the source of its production? When the three soil treatments listed above were used under a rotation with red clover turned under ahead of the corn which was followed the next year by wheat and this latter grain was fed, it was consumed (a) 100%, (b) 96%, and (c) 85%, respectively, for the order of fertilizer combination given. The complete treatment of lime, phosphate and potash was the hog's first choice in terms of the wheat grain so grown. But when sweet clover was the green manure ahead of the corn to be followed by the wheat, this second grain was chosen first when only lime and phosphate were used. The second choice was the more complete fertilizers, namely lime, phosphate and potash, and their third choice was lime only.

When a variety of different green manures was tested, each used separately, and with the same complete mineral addition given above, and when the corn grown immediately after the green manure had been turned under, was offered to the hogs, their choice of the corn grown immediately after these green manures ranked the soil treatments as follows: lespedeza 100%, red clover 90%, sweet clover 81% and timothy 66%.

In both of these citations given the hogs were seemingly "turning thumbs down" on the wheat and on the corn grains when sweet clover was used as a leguminous green manure either immediately ahead or one year ahead of the grain crop offered them. This was a case of the hogs seemingly voting against the sweet clover when it was used distinctly as a green manure.

Such was not the vote of the hogs however, when the corn followed a crop of sweet clover grown to maturity for the harvest of sweet clover as the crop rotation. On one plot the sweet clover was turned under ahead of the corn as a green manure to make this a two year rotation. On the other the sweet clover was allowed to mature and its seed was harvested. This left the sweet clover trash on the land during the third year in the rotation ahead of the corn.

With sweet clover as green manure and with (a) lime, (b) lime and phosphate, and (c) lime, phosphate and potash, to give increasing amounts of sweet clover to be turned under with the resulting more bushels of corn per acre as more fertilizers were put on the land, the hogs recorded their choices as follows: (a) 100%, (b) 80%, and (c) 67%, respectively for the above order of the fertilizers. They were not voting in favor of putting more sweet clover under by the help of more kinds of fertilizers, nor were they voting for more bushels of corn yield per acre. In fact they were "again" all of it. But with the sweet clover as a seed crop using the entire crop year ahead of the corn and with these same soil treatments the hog's vote was for the exactly reversed order of these soil treatments with the figures (a) 62%, (b) 85%, and (c) 100%, respectively.

In the case of this hog choice it was not the minerals added to the soil that corresponded with the particular selection of the grain by the hogs. Surely, then, it was not an increased ash content of the corn or of the wheat as a result of putting mineral fertilizers on the soil ahead of that particular crop that guided their selections. Rather it was some effects prompted by the nature of the organic material turned under as a green manure. When the hog, by its choice of the corn, votes against more kinds of fertilizer elements applied to the soil for sweet clover as a green manure under that corn, and then votes in the very reverse order when the sweet clover matures as a crop ahead of the same kind of corn under test, surely it is not the simple aspect of the mineral fertilizers as deliverers of such elements that exercises the effects recognized by the hog. Surely mineral fertilizers are not of influence only as minerals. Rather they are in control of what products the plants manufacture in consequence of the minerals' presence in the soil. Here by their most discriminating selections from amongst these different soil treatments, the hogs were demonstrating the same uncanny capacity as able compounders of their rations that they demonstrate before the self-feeder.

Now that we have been farming long enough in this country to have seen soft or starchy wheat follow the hard or protein-rich wheat going westward; and when the pork market centers have been following the beef market centers also moving in that direction, are we not about ready to believe that the exhaustion of the soil fertility is compelling the production of less protein in both the plant and animal crops and is limiting animal output more to mainly fat? When the wild animals roamed so widely and searched out the feed so carefully according to the soils growing it, can't we see that the fence that encloses the hog throws on us as husbandmen the responsibility of selecting the nutrients for the hog's ration with somewhat near the accuracy and delicacy with which the hog could do it herself? Have you ever thought that you might use the hog to test the corn lots from different parts of your farm or their difference in feed value? Are we not now approaching the time when perhaps an animal assay may be a good basis for discriminating against the use of one lot of corn from soil of neglected fertility and for acceptance of another lot of feed according to the hog's approval of the better fertility of the soil growing it? If your skepticism about this flares up why not let the hog demonstrate for you before the self-feeder in your own hog lot where you can try a few hundred pounds of each different grain and have the hog's judgment support you in the decision? When more hogs can speak to us by this method we shall not delay long our increased attention to the fertility of our soils for higher feed values. The conviction will soon become universal that more efficient hog production can well be founded on the fact that better soils make more and better hogs.

DISEASES AS DEFICIENCIES VIA THE SOIL

While it has long been common belief that disease is an infliction visited upon us from without, there is a growing recognition of its possible origin from within because of deficiencies and failures to nourish ourselves completely. Fuller knowledge of nutrition is revealing mounting numbers of cases of deficiency diseases.

These deficiencies need to be traced, not only to the supplies in the food and feed market where the family budget may provoke them, but a bit farther, and closer to their origin, namely the fertility of the soil, the point at which all agricultural production takes off.

These increasing cases classified as deficiencies are bolstering the truth of the old adage which told us that "To be well fed is to be healthy."

Isn't it good nutrition that is used as the "cure" for human tuberculosis? In that "disease" the effort is not given to the extermination of the microbes from the lungs and other body parts by means of antiseptics and other sterilizing agents.

Instead it is nutrition by milk, eggs, meat and all else for a high protein diet. Under such treatment, the germs apparently recognize their premature anticipation of a task of disposition and literally move out. Shall we emphasize the "cure" in this case or shall we raise the question whether deficient nutrition and defective physiology were in advance of, and an invitation to, the entrance by the microbes?

Were the "germs" the cause then, or merely an accompanying phenomenon of what is a deficiency but which we call tuberculosis? Might this not be the cause for some of our cattle diseases, accompanied by microbe, but yet so baffling that slaughter is still the "cure."

In cases of undiagnosable animal ailments, the able veterinarian often recommends feeding good alfalfa hay grown on the more fertile midwestern soils, or he prescribes some extra amounts of other protein supplements, as accompaniments to his medication. When the animal recovers, a similar confusion as to correct explanation of causes for the animal recovery is involved.

With the fattening of animals and its speculative aspects so prominent in agriculture, and with so little attention given to real production in terms of good nutrition because its failure is too commonly considered some "disease" or "bad luck," it will take some time before we appreciate fully the simple fact that the soil fertility is the foundation of the pyramid of all life.

We are slowly realizing that the soil is the source from which every branch in the assembly line of agricultural production is kept running full. As long as crop bulk and animals merely fattened for more weight are the major goals of our agricultural effort, our thinking to no greater depths will delay the day when we see the soil as support and in control of production.

That a soil may be speedily exploited of its protein producing power while its capacity for delivery of carbohydrate bulk holds on long afterward, is a potent fact that has not yet been recognized in our westward march. Under such circumstances we shall continue to talk about "buying" and rationing protein-supplements instead of accepting the costs of soil treatments to grow them.

When a soil is not fertile enough to make the protein in a seed crop, as was the case for early trials with soybeans, we say, "This is a hay crop, but not a seed crop." In that remark we show our lack of appreciation of the consequences for the poor cow asked to live, to reproduce, and to give milk while feeding on that hay. When such a "legume" hay was fed to fattening steers, it was a surprise that some of them "went down" on their hocks as if hamstrung and others with paralysis of the rear quarters.

We are gradually coming to see that "poor health" is creeping into the animals even while in the fattening process, because poor feeds result from poor crops, and the poor crops could do no better in their creative effort than was permitted by the soil from which alone creative potentials spring forth.

Unfortunately, for our domestic animals, it may take a goodly number of their disasters and deaths to convince us generally that much that is classed as animal diseases may be no more than nutritional deficiencies traceable to the low fertility of our soils growing the feeds.

WHAT'S NEW IN . . . SOIL HUSBANDRY

That we should think enough about the soil even to ask the question, "What's New In Soil Husbandry?" is in itself almost novel.

Not so long ago we did not include the soil when speaking broadly of husbandry. Only in recent years has it become a practice of folks in agriculture to consider production in terms of the fertility of the soil.

Crop production in general has moved more and more out of the whims of the weather. We are coming to realize that for a big crop both food and drink are required in ample supply.

Our fields are no longer just places where the crops are canopied under sunlit skies and sprinkled by rain. They now supply the fertility for crop creation. This is measured by detailed chemical soil tests and supplemented through test-guided soil treatments to give well-balanced plant nutrition, regardless of the variable weather.

As a consequence, rainfall is now being used more efficiently. What was once considered lack of moisture, like the yellowing of the lower leaves on cornstalks, is now known to be due to shortage in the soil's nutrient supplies, especially nitrogen. It is not due to a shortage of water.

Such newer concepts are bringing us to realize that probably the rainfall has been plenty. We probably have failed in soil husbandry to match it with the necessary plant nutrients.

Last season's high yields of corn, often more than 100 bushels per acre, were not due alone to the ample supply and distribution of the rainfall. They resulted where the grower tested his soil, and invested in calcium, phosphorus, potassium, nitrogen, and other nutrients necessary for the balanced nourishment of the increased number of cornstalks.

It was not a blind faith in the weather, nor in a particular kind or pedigree of seed. Instead, it was an effort to help the soil to really feed the plants. Farmers are beginning to realize that moisture probably has been plentiful. They have just begun now to build up the soil to the level of fertility necessary for its efficient use.

Recent studies by soil scientists and plant physiologists have stated that the deep rooting "habits" of any crop are more likely due to the greater depth of fertility in the soil than to any natural property of the crop.

Roots are growing bodies. They move through the soil in accordance with the fertility needed to feed them, rather than the pressure from the plant top.

Roots that are merely going through the soil are spindly and scarcely visible. They are literally in search of something that will feed them. Once they meet up with a center of more concentrated nourishment, like a clump of buried manure or a deposit of fertilizer, they suddenly become enlarged masses of big roots in that particular soil area.

For example, sweet clover has a spreading root system with many horizontal branches. In a shallow surface soil over compact, fertility-deficient, clay subsoil (Putnam silt loam, for example) it is shallow rooted. On another soil, like Marshall silt loam with good fertility distributed down through it, sweet clover is deep rooted.

It is the soil rather than the crop that controls the deep-rooting behaviors.

Now that antibiotics have demonstrated their benefits in human physiology, we are considering whether the physiology of all other life forms between microbes and man might not likewise get benefits.

These diverse chemical creations of the soil microbes taken along with their feeds by animals have been shown to make those rations more nourishing. Feeding some soil directly demonstrates its healthful effects of pigs.

For plants, too, we know of growth stimulating values, resembling those of chemical hormones, from rotted organic matter under careful tests.

Such thinking about microbes is decidedly new in soil husbandry. We are finally considering them as benefactors, when too long they were believed to be enemies. "Dirt" need no longer be held dangerous because of its "germs." Microscopic life of the soil is taking its deserved place in the biotic pyramid.

Located between the soil and the plants, microbes couple up with the soil as the starting point to give us the assembly line delivering the major chemical complexes of food values to man and beast at the top of the pyramid.

Newer theory and practice in soil husbandry emphasize the soil as chemical basis for our nourishment. Soil is not the source of only ash elements prominent in our cremation, but is the means by which microbes, plants, and animals chemically create or synthesize all the complex compounds required to feed us.

This is a newer line of thought in soil husbandry that may well be nursed by more research of science and by wider appreciation of the public as a whole.

7

NUTRITION

THEIR QUESTIONS—MY ANSWERS

QUESTION. Do vaccinations pollute the blood and do damage to the system?

ANSWER. If you take water into your system as a drink, it goes into your bloodstream directly from the stomach. But if you take fats they move into your lymphatic system. When you take other substances, like carbohydrates and protein, they go into the intestines and from there are passed through the liver, as the body's chemical censor, before they go into the blood and the circulation throughout the body. Most of your vaccination serums are proteins, and they are put by hypodermic needle directly into the blood and are not censored by the liver. Consequently vaccinations can be a terrific shock to the system. I try to avoid vaccinations. I react violently to some of them. When I had a typhoid fever shot for traveling abroad it made me deathly sick and left me stone deaf in one ear. I have only partially recovered my hearing in that ear.

QUESTION. Believe you said when you were a young man going to college the wheat in Kansas had as high as 17% protein and that it now is much lower.

ANSWER. Yes, the wheat values of Kansas as protein percentages in going from its eastern to western part ranged from ten to more than 18. By 1951 they ranged from nine to 14.

QUESTION. Should sodium propionate be used as a preservative?

ANSWER. It does not have much food value. Occasionally it sets up allergic reactions. Your body will probably handle it, but why handle excess? The liver is the only organ in your body that can regenerate itself. Why should we overload our bodies to knock out all too soon the chemical censor we have in them?

QUESTION. If you are just starting out with a little farm or garden, what, outside of manure, would be the proper supplement for the soil?

ANSWER. The first thing one should work out would be the proper calcium supply and likewise the magnesium supply. I would look after the calcium supply by the soil test. If you can't look after it by a soil test, you could grow some legumes and by their failure the need to build it up would be indicated.

QUESTION. What approach would be most effective in getting the facts to the government? (We presume the inquirer means the true facts about the organic or nutritional way of growing plants.)

ANSWER. You must learn that this business of studying nature takes a lot of hard work and late hours, and almost everyone is on an eight-hour day. We have lost our curiosity about nature. Unless one is curious, he doesn't learn much. I don't have much solution for you in connection with getting immediate results as to response by busy folks in government to what reforms some of us deem so necessary. It is usually about a generation before you get much encouragement.

QUESTION. Has there been any experiment in your type of agronomy in adding life to the compost heap so that it can multiply and grow?

ANSWER. In composting you have a kind of yeast performance with fermentation when the pile is moist. As the heating process dries out the water, air goes in, then an oxidation or burning process drives off carbon dioxide. In your surface soil you alternate those processes. Cultivation is the opening of the draft. "Energy" foods burn out of soil or compost. Growth foods are kept in by the microbial bodies. Then you have variations of synthesis by microbial growth in a soil of high energy supplies, due to the organic matter in it. When the organic matter of high energy, but low growth, value in the compost is spent, the synthesis is not dominant. The pile is cold. Decomposition sets in and that releases growth foods as well as energy foods.

Fungi have a tremendous faculty for drilling through tough organic wastes. The fungi and those rougher feeders starting the composting process have a greater power to break soil apart than do the plants. Bacteria, like plants, feed after the fungi have worked over the organic wastes.

QUESTION. In your soil tests of fertilizers I noticed that superphosphates were used.

ANSWER. When I went to Missouri 40 some years ago, there was one phosphate used, and that was superphosphate. At that time it was also recognized that phosphates were the major soil treatment with a good response by most of the farming crops. We are now appreciating the importance of phosphorus more in many biochemical reactions.

This year [1962] the man who won the Nobel Prize got it for photosynthesis, the mechanism in which he found phosphorus playing an important role. I have a feeling that phosphorus has a greater part in biochemistry than we realize. It is playing bigger roles in soils when combined with the organic matter there. We have phosphorus tied up in so many ways that we do not know all of its biochemical sources.

QUESTION. I am sure you are aware of the newspaper and magazine articles of the past several months, saying there is no difference between minerals that are organic and those that are not. Is there a difference between iron, sulfur, and others? Is there a difference between feeding plants organically, or organic minerals?

ANSWER. In your bloodstream you have iron in the hemoglobin. It is there in about the proportion of one part of ash to about 40 parts of combustible substances like carbon, hydrogen, nitrogen, or organic matter.

We have finally come around to the word "chelation." In hemoglobin the iron acts as part of the whole, not separated from it as sodium is in a solution of ordinary salt. In the hemoglobin we say the iron is "chelated." So we say phosphorus is chelated by soil organic matter and thereby more available to the plant from the soil. Nature doesn't separate the organic and inorganic.

QUESTION. My problem is this. I have a small city lot. I would like to grow my own vegetables. Is there some simple way, like using kelp?

ANSWER. You are talking about a biased case when you call on me. I use kelp. I am a hypothyroid case. I started using kelp about 20 years ago and I still use it when I start to get unusually tired whenever I feel I shouldn't.

This kelp is a good idea and it will take care of a large share of your other trace element needs as well as the iodine which serves for hypothyroid folks in many cases.

QUESTION. I am just a little confused on these superphosphates. I have always put organic matter on my soil, and I have had wonderful results. For organic matter I use cuttings, manure, oyster shells, or ground phosphates.

ANSWER. I would say you have just about covered the waterfront. You are putting pulverized phosphate minerals into compost. Thereby the phosphorus is more highly mobilized for plant nutrition. Manure supplies nitrogen, heating, and a whole series of changes. Oyster shells supply calcium, and you are completing the cycle of "out of the soil, back into the soil."

BALANCED SOIL FERTILITY— BETTER START OF LIFE

The agricultural use of limestone on soils is still mainly a practice based on empiricism; that is, upon experience or observation alone, without much theory put under scientific test. Since the early 1900s it has been contended that limestone (calcium carbonate) applied to the soil improves crop growth because the carbonate part of the compound lowers the degree of soil acidity. Various kinds of limestone, including that which is wholly calcareous or dolomitic in varied degrees and all those crushed to varying degrees of fineness, have been widely used.

Such stones have not been a costly commodity, for the finer lime-rock fractions have been a byproduct of the rock-crushing industry. The limestone fractions have served well for crop improvement, especially on soils in humid climates. This improvement of plant growth has not been widely attributed to the physiological or biochemical effects of calcium (and magnesium) resulting in better balanced soil fertility for both soil microbes and plants.

The neglect of basic research into the biochemical services rendered soil microbes and plants by calcium (and magnesium) may well be expected, since the farmer's income cannot underwrite it. Nor can the stone-crushing industry finance the necessary basic research, since the economics of so limited a manipulation of a naturally plentiful mineral—simply crushing and sieving—does not provide a great enough profit to allow the industry to underwrite research, much less set up its own laboratories.

Beginning with the year 1935 a federal subsidy (ACP assistance) for applying limestone to soils increased the practice (see chart below).

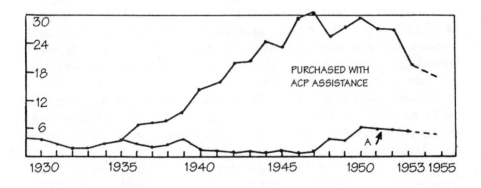

But even that was by no means enough to add the needed limestone over extensive areas. More recently, since federal subsidy, coupled with state funds, is building concrete highways which are utilizing the finer byproduct limestone, the stonecrushing industry has little interest in researching the use of limestone in agriculture. Our farmer's knowledge about research seems to be confined to what salesmen and their literature give him as "scientific" information.

Yet while one hears little about the application of limestone for calcium and magnesium for plant nutrition, much is still heard about nitrogen, phosphorus and potassium being used on the soil as commercial fertilizers. These are in the limelight for national support of younger nations in their struggle for freedom. Fertilizers carrying mainly these three nutrient elements, and new industries abroad to make only these, seem to be considered the major uplift factor in self-support through increased food production. One hears nothing about balancing fertilizer with the calcium (and magnesium) of limestone. We forget that calcium appears in vertebrate species in the largest amounts of the "ash" elements. We forget also that its deficiency in the soil is the major factor to be considered in classification of soil according to its productivity.

The importance of calcium in balance with phosphorus and potassium in the soil for feed crops was nutritionally demonstrated by Professor W.B. Nevens, et. al., of the Dairy Department of the University of Illinois,[1] when wheat grain and lespedeza hay (plus salt) were fed to New Zealand white rabbits. This ration grown on the same soil given fertility additions, differing in calcium only, demonstrated forcefully the importance of calcium in the soil. The soil with calcium produced crops which provided high nutrition for nursing mother rabbits and a better start in life for their offspring.

Dr. Nevens took advantage of the Newton Experiment Field of the Agronomy Department and fed the two crops grown in 1945, after fertilizer treatments had been continued since 1912. These treatments consisted of raw rock phosphate (a total of four tons per acre), kainit (2.25 tons per acre), and muriate of potassium (0.5 tons per acre) on each of a pair of plots. Also, on one of these limestone had been added (12 tons per acre) during the 33 years. The wheat grain and the lespedeza hay were grown on the plots with those soil treatments which are designated in this discussion by the symbol "PK" for the phosphorus and potassium treatments, and by "PK + Ca" for those treatments coupled with the limestone on the soil. Also the symbols "O" and "W" will be used to represent plot, animal, feed, or young, associated without and with limestone on the soil, respectively.

It is significant to note that the lespedeza hay crop "O" consisted of 33% lespedeza plants and 67% weeds. For the "W" hay crop the figures were 71.5% lespedeza and 28.5% weeds. The nitrogen content of the hay from the "O" plot was 1.96%; for the "W" plot it was 2.36%. The wheat grain from the "O" plot had a total nitrogen concentration of 1.53%; that from the "W" plot had 1.94%. The weeds were removed from the hay crop for the feeding trials.

So far, reports on the effects of limestone suggest more than can be readily explained by the uptake of calcium by the plant. The increase in lespedeza plants suggests that more seeds germinated and more seedlings survived because of added calcium in the soil. Dare we say that such results were due to lime as a neutralizer of soil acidity only? If so, limestone was only a "soil conditioner." But it has been shown that calcium applied at the time of seeding increases germination and plant stand (for clustered seeds like the sugar beet, in particular).[2] Calcium mobilizes more phosphorus into the lespedeza crop.[3] It also brings a higher percentage of the exchangeable potassium from the clay colloid into the crop.[4] These results are probably due to the fact that a higher degree of calcium saturation of the soil colloid modifies the root hair membrane to enable it to control nutrient in-go and out-go of the roothair in contact with the soil's organo-inorganic colloids.

The paired feeding technique was used to equalize the feed intakes. By this method, the amount of feed offered the pairs each day is equal to the smallest amount consumed by one of the two pairs on the preceding day.

Six pairs of litter-mate female white rabbits and two related males were used. After three weeks of preliminary feeding, the females were mated with the males. In most instances, the young were born at the end of the 31st day after conception. As soon as possible after birth, the young were counted and weighed. Usually by the third day after birth the litters were standardized to six young or less. These could not have access to the rations fed to the mothers, hence milk from the mothers was the only feed-source of the young.

Weights of the litters were taken daily until the 46th day after conception, that is, 15 days after birth, when the litters were sacrificed, individually weighed and measured for length. The gastro-intestinal contents were removed; the carcasses were weighed, frozen and later ground for chemical analysis as to (a) dry matter, (b) total nitrogen and (c) ash. In all, 46 carcasses were analyzed an equal number being from group "O" and group "W." Two litters were omitted since one doe on "O" ration died 35 days after conception, hence her litter mate "W" and young were omitted in making up the paired sets of data.

As indicated by the quantity of feed which remained in the feeders each day, the PW + Ca ration was slightly more palatable to the females than the PK ration. Four of the five on the PK ration limited the daily feed intakes of their pair mates on the PK + Ca ration.

Seventeen pairs of matings with the male by females on the two rations resulted in 90 young from 14 females on PK rations, or 6.43 individuals per litter; but in 93 young from 12 females on PK + Ca rations, or 7.75 per litter. Here calcium stands out for its creative power in the mere "handful of dust." While the conception rate of does on the PK + Ca ration was lower (12 out of 17) than for does on the PK ration (14 out of 17, yet the total number of young from the 12 does on the PK + Ca ration exceeded that from the 14, because of the higher fecundity resulting from the PK + Ca ration.

On each of the two rations, there were 23 young nursed by five does. The average daily weights of the litters per dam on the PK + Ca ration were larger than those of litters by dams of their paired mates on the PK ration. The differences in these weights were apparent at a week after birth, or 38 days after conception. These differences increased during the 15 days between birth and sacrifice.

The average body length for the 23 young from mothers on PK rations was 18.5 centimeters. For the 23 young from mothers on PK + Ca rations, the difference was 19.9 centimeters, and increase of 7.5%, due to the calcium factor.

The average weight of the stomach and intestinal contents of the 23 young resulting from PK rations was 12.6 grams. The corresponding weight for the same number of young from PK + Ca rations was 16.8 grams, an increase of 33%.

The chemical contents of the carcasses as (a) dry matter, (b) total nitrogen and (c) ash, all in terms of average weights in grams for PK + Ca rations, were (a) 39.66, (b) 4.42 and (c) 5.36; but for PK rations they were (a) 35.36, (b) 3.86, and (c) 4.79. This shows the larger amount of synthetic or creative organic results from the ash contributions of the soil when it was balanced by calcium, or limestone.

Considering only the percentage compositions as measure of difference between the results from PK + Ca ration and the PK one, the former ration was represented by (a) 24.65%, (b) 2.75% and (c) 3.39%; the latter ration was represented by (a) 25.52%, (b) 2.78% and (c) 3.51%. The calcium in the soil did not modify significantly the percentage concentration of (a) dry matter, (b) total nitrogen or (c) ash of the carcass or body. Its effect was to increase the weight of body tissue, not to change its chemical composition.

Calcium is powerful in building more living tissue or in helping to bring into play other nutritional factors. Accordingly it is valuable for its service in "amending" the soil rather than merely as a fertilizer to increase plant growth.

All of this tells us that we are not yet fully familiar with the biological functions which the soil-borne elements perform in plants and in animal consuming them. Much less do we understand the role of these elements in the mother's function of multiplying the species and sending them forth to be healthy and to populate the earth. We are slow to study the importance of soil fertility to the quality of food, for this is not yet to our economic advantage in the marketplace.

SOIL FERTILITY AND THE HUMAN SPECIES

We have often been told that "armies march on their stomachs." We have not been told that stomachs march according to the fertility of the soil. The fertility of the soil consists of those chemical elements of rock origin that contribute to the construction and body functions of both plants and animals.

Origin or Source	Essential Elements	Human Body %	Vegetation % Dry Matter	Soil %Dry Matter	
			Table 1		
		Chemical Analysis of the Human Body in Comparison With That of Plants and of Soils.			
Air and Water	Oxygen	66.0	42.9	47.3	
	Carbon	17.5	44.3	.19	
	Hydrogen	10.2*	6.1*	.22*	
	Nitrogen	2.4*	1.62*	—	
		96.1%	94.92%		
Soil	Calcium	1.6*	.62*	0.3‡	3.47*
	Phosphorus	.9*	.56*	0.0075	.12*
	Potassium	.4†	1.68†	0.03	2.46†
	Sodium	.3	.43	—	
	Chlorine	.3	.22	.06	
	Sulfur	.2	.37	.12	
	Magnesium	.05	.38	2.24	
	Iron	.004	.04	4.50	
	Iodine		Trace		
	Fluorine		Trace	.10	
	Silicon	Trace	0-3.00	27.74	
	Manganese		Trace	.08	

* These are involved in the plant and animal struggles to find enough to meet the high concentrations needed.
‡ Amounts common as the more available forms in the soil in contrast to the total, most of which is but slowly available.
† This represents struggles by the animals to eliminate it.

Let us, at the outset, deal with soil fertility and the human species on seemingly low levels. We can consider man mainly as a physiological performance. Let us, however, not be limited to an exhibition of digestive processes only, and let us entertain the hope that man, in good nutrition, balanced physiology, and the resultant good health, may manifest his appreciation of, and respect for, life and that he may demonstrate the finer attitude and the nobler spirit. It is along this line that soil fertility operates to determine the nature of the human species.

The human species may be reduced to a very simple basis. Man is about 5% soil, or 5% ash. This represents the soil's contribution to the construction of the body. The list of elements coming from the soil includes calcium, which makes up 1.6% of the normal body weight. Thus in a body of 150 pounds there is the equivalent to slaked lime required to lay a half dozen bricks. The next element is phosphorus. This makes up 0.9%, or about one-third pound per adult. The other elements of soil origin come in the following order: potassium 0.4, sodium 0.3, chlorine 0.3, sulfur 0.2, magnesium 0.05, and iron 0.004%. There are traces of iodine, fluorine, silica, manganese, and others.

Plants require the same elements as animals, except that animals demand two, and possibly a few more, of which the essentiality is still in doubt. Calcium and phosphorus head the list in animals and in plants, except for potassium. Calcium, as a part of the ash, represents increased concentrations from 1, in the soil to 8, and to 40 in the plant and animal ash, respectively. For phosphorus, the corresponding figures are 1, 140, and 200. These increased concentrations represent a physiological struggle by an animal body to maintain them, or to substitute for them, under the threat of disturbed or perverted body form and functions. In the case of potassium, the animal struggle is apparently not one of sufficient supply, but of elimination, if one dare infer from the high concentrations in plant but the low animal body concentrations; the commonly pronounced excretion in the urine; and the uncertainty as to its exact body functions.

An adult body contains but about 7.5 pounds of soil elements. Life without any one of them becomes impossible. In terms of the plant that demands these ten from the soil, the absence of any one prohibits the plant from functioning as the greatest of all chemical industries. Some micronutrients, demanded in only parts per billion, are requisite for this great out-of-door manufacturing business, where water as the hydrate from the rainfall, and carbon from the atmosphere, are synthesized by the sun's energy into carbohydrates, and these in turn, by micro- and macroscopic plants further synthesized into proteins and other complexes.

The meteorological aspect of plant growth crowds itself to our attention, when carbon, hydrogen, oxygen, and nitrogen, coming from air and water and not from the soil, make up almost 95% of plant and animal bulk. Here is basis for the prevalent but erroneous belief that weather is in control of crops. We need seriously to consider that growth occurs more according to the soil and less according to the weather. We need to look into the possibility that by supplying fertility to the soil we can mitigate the hazards of the weather.

The small portion of nutrition supplied by the soil puts it in control more than we are wont to believe.

Plants, like humans, can be said to be—and to behave—according as they are nourished. "How can this be true," you may well inquire, "when it has been the common concept that climate determines what kind of plant is grown and what kind can be grown?" To give full validity to this newer concept is to emphasize the indirect more than the direct control by weather over life forms. We quickly see our body comfort in relation to the heat and the cold, which two, within the less general limits, are not the complete determiners of plants.

For plants, the nutritional aspect—another significant human comfort—is truly in control. Plants are distributed over the surface of the earth according as they are nourished to provide the particular chemical composition which they, as species, represent. Climate is involved in this nutrition, but more as it is a factor in determining the soil. The soil in turn is determined, not only by the climate, but also by other factors, such as the rock, or the parent material, which is converted into the soil by the climatic forces. Climate alone is not in control of plants directly, but jointly with other forces working through the soil, and therefore indirectly.

In accordance with the long-held belief that only meteorological conditions determine the plants in any locality and their adaption to the environment, we have been scouring the world and making transplants with little regard for the soil fertility required to nourish the shifted plants. When alfalfa grows dominantly in Colorado soils; when sugarcane grows abundantly in Louisiana; and when the rubber tree quickly dominates in tropical Brazil; are these merely matters of differences in the degrees of temperature or of the increasing amounts of rainfall? There are, of course, meteorological differences, but are these alone the causes? Can plants be shifted merely by keeping them properly heated or properly moistened?

Alfalfa is a protein-rich, mineral-rich forage of high calcium content. It demands large supplies of mobile nutrients from the soil. It grows well where lesser amounts of rainfall have not depleted the lime and other mineral fertility stores from the surface soil. When planted on a more humid soil it demands particular soil treatment for its successful growth.

Cotton is a deliverer of products that are mainly carbonaceous, represented by its cellulosic fiber, oily seed, and shrub-like form. It demands less of these fertility elements requisite for alfalfa. Lime application to the soil is not an absolute requisite for cotton, though the crop is improved by it. Cotton responds more to potassium, the nutrient associated with carbohydrate production in plants. Cotton is not on the list of the vegetations that serve as animal forage. This is because its composition is such that only a limited part has food

Table 2

*Range in Fluctuation of Concentrations of Chemical Elements in Forages.**

		Ash %	K_2O %	CaO %	MgO %	Fe_2O_3 %	P_2O_5 %	SO_3 %	SiO_2 %
Mixed meadow	(Low	2.02	7.60	6.00	1.90	.10	2.00	.70	10.40
grasses	(High	11.40	56.60	40.10	24.40	4.90	21.30	13.40	63.20
	Range	9.38	49.00	34.10	22.50	4.80	19.30	12.70	52.80
Red Clover	(Low	4.50	8.80	21.90	5.30	.30	4.00	1.20	00.00
	(High	9.20	52.00	53.40	26.10	5.00	15.00	7.40	20.20
	Range	4.70	43.20	31.50	20.80	4.70	11.00	6.20	20.20
Alfalfa	(Low	5.40	11.40	24.70	2.80	.50	4.50	3.70	.80
	(High	9.50	41.90	62.90	9.00	8.20	19.30	8.60	27.90
	Range	4.10	30.50	38.20	6.20	7.70	14.80	4.90	27.10

* USDA Yearbook 1938. C.A. Browne. Some relationships of soil to plant
and animal nutrition—The Major Elements, p. 781

value for animals and then mainly after processing.

The rubber tree is a woody crop. Its latex is indigestible. Like other forest trees, it uses soil fertility for growth, but each annual supply is dropped almost wholly in its leaf crop. Through decomposition, this fertility supply in the leaves completes its regular cycle as it rotates from the soil up through the tree to the leaves and from the fallen leaves back to the soil. While making this cycle it is building the latex and the wood. Both of these are made mainly of air and water elaborated by sunshine into energy compounds of fuel value only for flames and not for the physiology of animals.

"A series of plants like alfalfa, cotton, and rubber trees," it may be thought, "fits well into the climatic variations." But it is well to note that animals fit into this picture too, but more in terms of nutrition than their body comfort. Then the controlling force is not the climate so much as the soil fertility that determines the composition of plants, and thereby their nutritional services to animals. Alfalfa nourishes animals, but cotton plants and forest trees do not. Evolutionary forces located the plants in correspondence with the soil fertility. Soil fertility is operating over wide geographical areas to give the particular ecological array of plant species in which each species has a chemical composition according to the level of soil fertility. Since the individual is an epitome of the species there is a variation within individual plants in chemical composition, according to the variation in the soils over which that species survives. As the kind of forage is one that falls lower in concentration of nutrients, apparently it is likewise one that tolerates wider fluctuations in chemical composition as table 2 illustrates. This is a basic principle of greater importance in terms of our national health than we have been wont to believe.

Nutritious herbages and dense animal populations dominate in regions of lower rainfalls and moderate temperatures because the less weathered soils of higher fertility make for better nutrition. Such domination is not wholly true because of the comforts of climate. Animal populations in the depleted soils with forest vegetation are scant regardless of moderate temperatures, as the few turkeys found by the Pilgrim fathers testified. Soil fertility for vegetation of mainly fuel value, as cotton or forests, does not guarantee rapid animal multiplication. On the prairies in regions of lower rainfall and on more fertile soil, bison were numerous. The plant composition of high nutritive value reflects the higher soil fertility. As the soil fertility declines, the particular plant species in prevalence shift to those making more use of air, water, and sunshine in their body construction. Such plants can possibly support fattening animals, but are of doubtful value for young and growing animals. Beef calves are produced on the mineral-rich, growth-promoting soils of the drier range, but are fattened on the feeds of fuel value from the humid soils of the cornbelt.

The soil fertility on an individual farm can be depleted enough through failure to return manure, crop residues, and other fertility forms in a single human generation, to shift that farm from a place of good health to one of deficiency diseases for the farm animals and for the families on it. The same crops, still growing after 50 years of farming, may have shifted from protein producing, mineral-supplying, health-giving sustenance to vegetation mainly of fuel value, with nutrient deficiencies. The shifts may occur without changes in tonnage output. Here is a national weakness that is being heralded by a loud voice—but apparently going unheard—in the rejection figures, for example, of the Army draftees. We, as higher animals, along with the lower ones, are experiencing increasing nutrient deficiencies because the declining soil fertility is giving us food that is mainly of fuel value. This shift to foods mainly of fuel value is aggravated still further by processing methods in which the starches and sweets are retained and the minerals discarded. The shift is undermining reproduction and other delicate body functions. We are about to appreciate the fact that our soil fertility is the place where we may undergird rather than continue to undermine the national health.

Viewed in simple geochemical fashion, soils are ephemeral. They are rocks in various stages of progress in going from mountain to sea or from soil to solution. Silicic acid is passing out as it bows reverently to the quiet, but persistent, onslaughts by the simple and weak carbonic acid. Soils are the many intermediate products during this change from the silicates of the basic elements to their nitrates, phosphates, sulfates, extensive carbonates, and other simple products.

Reduced to a simple scheme, the processes of soil formation and development, or this march by rock to the sea, may be divided into two stages. The first is mainly construction in which clay and organic matter increase in the soil. There is also an increasing capacity and content of nutrients of service in plant and animal life. This occurs because colloidal adsorption and colloidal exchange of plant nutrients come into more prominence as the soil's content in clay and humus increases. These two, the humus and more particularly the clay with silica in dominance, are the constituents that carry increasingly larger stores of soil fertility. These stores include calcium, magnesium, potassium, and other elements in adsorbed forms, not leached readily by water, yet exchangeable to plant roots. These are the soils in which calcium dominates over all the other nutrients.

The second stage in soil development is destructive. Increasing climatic forces, in forms of heavier rainfall and higher temperatures, give the soils a higher clay content. This clay gives up its adsorbed calcium, very rapidly while its magnesium and other bases are less rapidly exchanged for hydrogen, through leaching and removal by plant growth, to become an acid soil. Under still higher temperatures and more rainfall, the clay itself increases in quantity but is also changed in chemical nature through reduced dominance by silica. As a consequence, such clay no longer retains plant nutrients or bases readily in the adsorbed forms. It no longer holds hydrogen to make the soil significantly acid. It is then neither acid nor loaded with elements of soil fertility, but is chemically inert. This inclination of the more highly leached clay toward neutrality—or what is more properly a chemical indifference because the clay does not hold even hydrogen—has been interpreted by many to mean that the soil has no need for calcium. Quite the contrary, such soils are decidedly deficient in this element.

The United States is divided according to this plant of soil development into two main areas. Starting in the western arid states and coming eastward and northward, soils illustrating the process of increasing construction and of increasing content of clay and organic matter are met. The maximum of this organic matter and of dark color, so commonly associated with high fertility, is reached in the territory of the Red River Valley in the Dakotas and Minnesota. From this point southeastward to Florida one moves through soils with increasingly weathered conditions to meet, first, the acid soils, those depleted of their fertility by cropping and weathering to give hydrogen substitution for the nutrients, and then, the more reddish soils where the clay has taken on a different chemical nature giving less acidity.

This different nature of the clay is responsible for the properties of the southern soils that underlie economic and social conditions of the people living on them. These conditions cannot be remedied by means of legislative enactments or political practices and procedures. The problems of the South tell us that the South needs soil fertility.

Botanists and soils men think of plants as getting their chemical nutrients from solutions. We know that the nutrients in true solution, or in the common ionic forms, are of such low amounts within the soil that limiting a crop to this supply would barely start the plants. Nutrients in the soil serving for crop growth are not in true solution, but are adsorbed on the colloidal clay and the humus. They are taken by the crop only as the plants can offer something in exchange or in trade. This offering by the healthy, luxuriantly growing plant is mainly the hydrogen ions. The composition of the clay according to the degree of climatic development of the soil, and in terms of the concept that the plant is a system of chemical equilibrium between root and clay in contact, to be shifted by exchange from the clay to the plant and vice versa, explains much more clearly the chemical composition of the crop in relation to the climatic region producing it.

The clay alone serves as an intermediary merely passing on what it gets. It cannot continue to produce crops indefinitely. This exchangeable nutrient supply on the clay and the humus, commonly spoken of as the colloidal complex, is readily exhausted by one, two, or three crops. There must be activities, therefore, in the soil to renew the exchangeable supplies if crop production is to continue. This renewal is brought about through organic matter decay and through mineral breakdown. Such breakdown occurs because the hydrogen put on the clay by the plant serves to exchange itself for other cations of nutrient value in the mineral crystals and rock fragments of the soil.

These advances in our knowledge about soil processes carry with them the concept that soil acidity is not detrimental to crop production because of the incidence of the acidity, or the hydrogen ion. Rather, the detriment by soil acidity results because of the exit from the soil of the fertility ions replaced by the hydrogen. Excessive soil acidity is merely a pronounced depletion of the plant nutrient supply from the clay. It represents an extensive exchange by hydrogen in the carbonic acid from the roots, or from percolating waters, for the plant nutrients on the colloidal part of the soil. It indicates, too, a deficiency in mineral nutrient reserves that would buffer this excessive clay depletion. As acidity increases, or more specifically as the fertility declines, through hydrogen exchange for it, the growing crop shift as to species of differing composition and chemical composition within the single species. This shift in their chemical composition is from the more proteinaceous products of high mineral content to those which represent more purely carbonaceous products, or a shift from alfalfa to cotton and to rubber trees and other forest vegetation on highly developed soils.

If an agricultural crop cannot starvingly accommodate itself far enough to meet this change in chemical composition and fails to grow, a different crop species is substituted. If that makes tons where its predecessor failed, it must have more of the carbonaceous makeup. Thus the nutritional bases for our plants have been slipping to lower levels. The plants, in turn, are giving forages of lower nutritional values to animals. The human species, likewise, is moving to lower levels, though with significant lag in the process.

There is evidence that the soil offers relatively much more soil fertility to the plant growing in the spring. Plants at that time deliver calcium, phosphorus, and other minerals in the forage at higher concentrations. This is because the lower temperatures and the less intense sunshine are giving lower rates of carbohydrate production, while there is a high mineral intake from the soil.

In our crop "juggling" the so-called new and imported crops have been hailed as successes. This has rested on their ability to take less from the soil but more from the air, water, and sunshine above the soil, and to hide more cleverly their lower value as forage feeds than the crops they have replaced.

Our soil exhaustion, or our fertility depletion, not only has encouraged our extensive shifts in kinds of crops, it has moved us from emphasis on grain crops to emphasis on forage values. Grains are more nearly constant in nutrient composition and reflect declining fertility by declining yields as bushels per acre. Forages tolerate extreme ranges in composition responsible for many animal irregularities. Nurse crops have almost disappeared. We now use sequences of crops in the same year instead of nurse crops, as in barley and lespedeza when barley grows in the fall and the lespedeza follows the next summer. The fertility delivery rate is too low to support the nurse crop and start the legume crop simultaneously. "Plant diseases" have increased. In reality they ought to be called "symptoms of plant starvation." Seed germinations are lower because we have never believed that soil fertility played a role so early in the plant's life.

Animal ailments are also on the increase because of the declining soil fertility. But as yet no one has tabulated them by localities according to soils of different degrees of exhaustion. Acetonemia in pregnant milk cows, milk fever after calving, pregnancy diseases in sheep, contagious abortion in cattle, rickets in young animals, and numerous other ailments still baffling as to physiological explanation do not occur in June when the animals have had opportunity to get from young grass the more concentrated forms and larger amounts of what we call soil fertility. Calcium gluconate as venous injections in the cases of milk fever and acetonemia cannot be substituted by the corresponding sodium compound. Calcium and phosphorus in particular are concerned in these irregularities. It should not be an impossible stretch of the imagination to connect these chemical elements so common as soil treatments for better plant growth with the physiology of foetus building or the production of milk.

Animals have been pushed to the dangerous precipice until decreased reproduction, increased diseases, more body malformations, and other irregularities have compelled us to market these animals early. What the use of this meat has been doing to human health has not yet been given consideration.

Where do humans fit into this soil fertility picture? Keen minds among the doctors of medicine and of dentistry, with greater desires to serve in prevention than in giving only relief, have seen degeneration in bodies, minds, and souls taking place at the highest rates among our peoples claiming the maximum of knowledge, invention, and standards of living. There is an increasing number of those who, with Ernest A. Hooton, of Harvard University, believe that we should be "finding out what man is like biologically when he does not need a doctor, in order to further ascertain what he should be like after the doctor has finished with him. It is a very myopic medical science which works backward from the morgue rather than forward from the cradle." To supplement this quotation, we should be working forward not only from the cradle but even from conception and preconception to give maximum of health to individuals.

Studies by such men as Dr. Price of Ohio, Dr. Pottenger of California, and Dr. Forman of Ohio, make it possible to see human deformities associated with nutrition, and this nutrition going back to the crop, the season, and the soil itself. It has now become hopeful to link man, in spite of his nomadic wanderings or his nourishment from far-flung food resources, to the soil and to soil fertility.

Calcium and phosphorus play no small role in the skeleton, and in the functions through which the skeleton acts as a contributor to shortages or as a depository for reserves. Probably other nutrient elements will fit into the picture. Unfortunately, national attention has not been developed to the point of recognizing the possibilities in putting these fertility items back into the soil and setting in motion the prevention of degeneration.

Declining supplies of calcium, phosphorus, and other items to handicap the crops are at the basis of our degeneration. Diseases are astoundingly on the increase. They are also on the relative increase. Hay fever, unknown to the Indian, has doubled its percentage of the population in 25 years. Heart disease, according to the New York City Health Department, took 203 per 100,000 in

1907, but 327 in 1936, a 60% increase in almost 20 years. Arthritis is on the increase. Cancer is on the increase. Dental caries, ailment of the exposed or visible part of the skeleton, is on the increase.

That the hidden part of the skeleton should be undergoing decay corresponding to that of the teeth is possible. Dr. Hooton of Harvard says:

". . . degeneration tendencies have manifested themselves in modern man to such an extent that our jaws are too small for the teeth they are supposed to accommodate. Let us go to the ignorant savage, consider his ways of eating and be wise. Let us cease pretending that toothbrushes and toothpaste are any more important than shoe brushes and shoe polish. It is store food that has given us store teeth."

Dr. Price, in his study of the teeth of savages in relation to diet, has made extensive chemical analyses of their foods to support his numerous oral examinations among primitives showing their almost perfect teeth. He reaches the conclusion that our narrowed and shortened lower jaws and narrowed upper jaws with displaced teeth, represent a compress on the lower brain to bring pressure on the pituitary and to set in motion a chain of mental and physiological disturbances. Narrowed faces, narrowed bodies in that "slim trim boyish figure," link themselves with our tendencies to refine foods "to make them sources of energy without normal body-building and repairing qualities."

Recent studies about the soil in which the effects of the soil treatment were measured in terms of different growth responses by animals consuming the forages grown on them, bring forth the possibility of feeding our animals by treating the soil. Calcium and phosphorus, the more common deficiencies in soil fertility, supplied to the soil, served to make each acre from 50 to 100% more efficient in growing animals. Balancing the soil fertility as diet for the plants increased the plant growth and the acreage yield. Each unit weight of crop was of distinctly higher value as balanced diet for the growing animals.

Improved physiology of the animals was also manifested and body functions operated more efficiently. Skeletal consumption and depletion processes centering particularly about the calcium and the phosphorus level came into prominence. Rickets was induced by feeds from untreated soils but was prevented by feeds from treated soils. Supplementing feed from the untreated soil by means of calcium and phosphorus as minerals was not the equivalent of using feeds grown on treated soil. This fact suggests that calcium and phosphorus as minerals was not the equivalent of using feeds grown on treated soil. This fact suggests that calcium and phosphorus coming by way of the plant growth processes do more for animal nutrition than merely deliver themselves to the animal's digestive system.

When only a preliminary trial of two of the major fertility elements, calcium and phosphorus, as soil treatments offers so much in improved body functions, and when other fertility elements are equally as simple or more so, there should be significant optimism ahead. When chemical attention to a more refined degree goes to the soil and plants, and when the information by the soil chemist and the plant physiologist as plant nutrition interrelates itself with that mass of information by the nutritionist, the manifold ramifications of soil fertility into the behaviors of the human species will be revealed as a great service by way of attention to the soil.

Let us hope that knowledge of our soil will arrive before human nutrition goes so low through neglect of soil fertility as to reduce thinking capacities to the point where we cannot save ourselves by saving our soils.

BLOOD WILL TELL

Our soils are no longer maintaining what for them might be called a "strong constitution" either, if we mean the animal's ability to remain free from diseases. The increase in new and unnamed animal afflictions is supporting that statement. We are not ready to believe that "blood will tell," and will report health irregularity more delicately than it is recorded in commonly visible symptoms.

When the cow takes the forages (also the soil and the microbes on them) into her paunch where this combination nourishes the microbes first, is it too great a stretch of the imagination to see her body's nutrition as a close connection with the fertility of the soil? She initiates her digestive processes by help of the microbial content in the fore part of her alimentary canal. The products of the transformations in the organic matter by the microbes, the digestive results of the cow's activities and the protein of the microbial cells so profusely multiplied, are all finally digested and absorbed into the bloodstream for its reactions in summation. There is in it the reflection by a warm-blooded animal of what the soil offers as fertility and as nutrition. Here then is the suggestion that the close chemical study of the cow's blood will tell not only the cow's condition as health, but may even point out the deficiencies in the soil's fertility responsible for the health deficiencies.

That the health of cows is so closely connected with the fertility of the soil might well have been the theory set up for test and demonstration in some epochal studies by Doctors J. Stewart (recently deceased) of Edinburgh and J.W.S. Reith of Aberdeen, Scotland. The experiments were made on a dairy farm plagued each early spring with attacks of "lactation" tetany or "grass" tetany. It was these studies of the increase, during the growing season, of the concentration of magnesium in the pasture herbage and the more common occurrence of tetany with generally disastrous results at the lower magnesium values, that brought the scientists to use a herd where tetany was common in the spring. They examined the cows' blood for concentrations of magnesium before treating the pasture soils with dolomitic limestones as magnesium fertilizers. Then for four years they followed the variations of this nutrient element in the herbages and in the blood, during and after the disappearance of the previously common afflictions of the cows with what was believed to be "grass" tetany.

They found that the concentration of magnesium in the herbage is very low in April, May and June and increases from 60 to 100% by August, to remain nearly constant thereafter. The common type of tetany coincides with the part of the season when the concentration of magnesium in the pasture grasses is at its lowest. The scientists also found that "magnesium limestone dressing increased the magnesium content of mixed herbage throughout the growing season by at least 50% above that of herbage receiving no lime dressings or ordinary limestone dressings. Such an increase in the magnesium content of pastures in April and May will increase the magnesium intake of the animal grazing them by about 20 grams per day, assuming that a dairy cow will eat 30 pounds of dry matter per day and that the pastures' content of magnesium oxide has increased from 0.30 to 0.450%.

"This is comparable to the amount of magnesium each cow obtains when fed five ounces per day of a high-magnesium mineral mixture containing 16% magnesium as suggested by Blakemore and Steward (1934) and shown by them to reduce the . . . risk of tetany in herds put to pasture in April or May."

The study of the blood of the cows showed that in place of the normal levels for magnesium at 2.5 to 3.0 mgs. per 100 ml. of serum with 2.0 as the lower limit of normal range, the mean values in April, June, and July were 1.71, 2.38 and 2.56 mg. respectively, before the soils were treated on the farm where tetany had been common trouble. The following spring two long tons of dolomitic limestone containing 41.8% magnesium carbonate and 55.30% calcium carbonate were applied on one part of the pasture. Another part was left untreated. A third part was given the same rate of a nearly pure calcium carbonate limestone.

Their successive seasonal tests, for the years after the soil treatment, showed the magnesium in the herbage in April and May at the low mean values of 0.17 and 0.16% of the dry matter, respectively, for the no-treatment area; of 0.19 and 0.17 respectively for the area given calcium carbonate stone; and of 0.30 and 0.26%, respectively, for the herbage went higher with the advancing season. By October and November, the figures were 0.26 and 0.26% respectively for those months for the plot given calcium carbonate only; and 0.37 and 0.40% for the plot given the dolomitic limestone.

Herbages on three other farms also suffering regularly from tetany were included in the study. The magnesium contents of their herbages in April, May and June were at the lows of 0.12, 0.13, and 0.17%, respectively, with the August, September and October values at 0.26, 0.34, and 0.30%. These were in close agreement with no-treatment or calcium carbonate treated values on the farm where the magnesium values of the blood of a dozen cows were studied for four years in order to follow the report by the cow's blood telling of the dolomitic soil treatment reflected by higher magnesium levels in this circulating form of body protein.

IS SOIL FERTILITY VIA FOOD QUALITY
REPORTED IN YOUR VARIED PULSE RATE?

"When are you allergic?" has been about as debatable a question as "when are you drunk?" Refined chemical tests of the blood have recently come into the legal domain, to distinguish differing degrees of the latter disturbing condition of your body in relation to what you drink. Now the simple counting of your heartbeats, in relation to what you eat, has been cataloged to detect how seriously you are allergic. Even alcoholism and allergies may have some closer connections and similarities, as the studies and reports of Dr. Coca in his recent book point out.

This publication of 189 pages by an M.D. deals with prevention, and thereby your own cure, of allergies. To him these do not mean only hay fever, hives, and asthma. They include epileptic seizures, high blood pressure, stomach ulcers, diabetes, eczema, colds, migraine headaches, hypertension, and a host of other commonly tormenting ailments which do not submit to medication.

Dr. Coca is the conceiver of the particular premise and its establisher as a basis of health—namely, "your pulse rate is accelerated by the foods and substances to which you are allergic." He pioneered the simple, safe, and harmless pulse test by which one can discover his allergenic foods. Then, by the avoidance of them, one can eliminate his own ailments. These are usually in numbers far greater than on the commonly considered list of allergies.

This pulse test suggests itself as a new but simple technique by which one's own eating, in relation to the subsequent changes in rate of heartbeat, can measure and report the quality of his food, even as the soil where it was grown determines it.

It suggests that a gardener and his family can play guinea pigs and run regular bio-assays of his soil management and treatments for their own better health. By giving only a part of the garden a particular soil treatment, and by planting the rows of vegetables across both this and the untreated soil area with seed from the same source, the eating of the vegetables from each treatment separately—followed by the pulse counts and the data therefrom—can give the possible allergic effects coming via the soil in the food differences.

Here is a simple means of bio-assay by the human body itself, rather than by guinea pigs or rabbits, of the nutritional quality of the vegetables according to the fertility of the soils growing them. One need not be a chemist, nor a doctor of medicine. You need only to adopt the practice of counting pulse beats in the morning before rising, then before and after eating the vegetables, to appreciate fully the meaning of the old adage which tells us that, "The proof of the pudding is in the eating thereof."

Dr. Coca is also an immunologist, a bacteriologist, a chemist, a clinical professor of medicine, a former medical director of the Lederle Laboratories for nearly a score of years; he is a teacher, a member of many learned societies, author of several scientific publications in the field of allergies, and a contributor to medical journals in many parts of the world. His book, *The Pulse Rate*, is written in plain and simple layman's language reporting a significant discovery of health via foods.

It gives the details of the method of putting it into practice by anyone who cares to observe and study the reactions of his body to the foods, etc., he eats—as they hinder or help his health. It bids fair to make each of us more nearly the captain of our own well-being through buoyant health.

HIGH TIME TO LEARN ABOUT
OUR SOILS AND OUR HEALTH

We are beginning to be concerned about our national needs for food, fiber, and shelter. No longer taking our resources for granted, they are being catalogued as consumable, renewable, scarce, abundant, etc. We see our population mounting at geometric rates, while resources per person are becoming less and less.

Little attention is given to the possible long-time deficiencies recognizable only when they approach disaster. Food, as it has been losing its nutritional quality while agriculture becomes a fixed one in place of one that was nomadic, has not yet come in for consideration of its nutritional deceptiveness. Nor has the fertility of the soil growing our foods and our feeds been considered as significant in the health of ourselves and our animals. We have been content with postmortems. We have been satisfied to work backward from the grave and the morgue in our search for health. We are slowly coming to approach health from the ground up, from the quality of the food as the soil growing it guarantees the inorganic elements, the vitamins, and the proteins as tissue builders as well as the carbohydrates for only calories and fuel values.

The pioneer farmed to feed himself and his family from the farm and not from the grocery store or the drugstore. For him "to be well-fed was to be healthy," not highly fattened. Modern farmers farm for economic reasons, for profits, for dollar values, and not for nutritional values. Modern agriculture views itself through the eyes of the industrialist who converts and transforms materials. These he assumes to be available. While agriculture may convert its products, it is not mainly a technology and that alone. It is first biology and technology second. It deals not in lifeless materials. It is concerned with living matters. It promotes the processes of creation, all originating in the soil. Soil depletion, by taking the soil for granted as if agriculture were only a mining industry, has brought us face to face with present shortages of the proteins.

Quality of food and feed are declining under the criterion of agricultural production mainly for sale. With yields per acre measured only in bushels and tons we have dropped out more of the protein-producing crops and have brought in the carbohydrate producers. Starches, sugars, fibers have increased but proteins and all the other body-building, and body-protecting qualities have decreased. Juggling the crops with substitutes for those starved out on declining soil fertility has given us those which are "hay crops but not seed crops." We are accepting those which can't make seed to reproduce themselves. Going to a grass agriculture appears as an escape from seed production but casts doubt on the possibility of the cow's surviving by it.

We are worshipping calories, while the proteins are still too crude to be complete nutrition. Supplements of protein are slipping out too. Corn is deficient in the essential amino acids, namely tryptophane, lysine and methionine, but its yields per acre for sale were pushed up by hybrid vigor. Its proteins slumped tremendously.

Failing reproduction goes with proteins failing in quantity and failing in quality. Missouri's pig crop marketed is only 60% of those given us by sows in their litters. The Missouri dairy calf crop at weaning time is only 60% of the conceptions. Putting the blame on "diseases" and killing the cow to escape them, seems an absurd approach by legal minded veterinarians to troubles in cows' or pigs' health going back to the soil fertility for prevention.

The animals have survived in spite of us and not because of us. We are keeping livestock because we keep them close to their birthdays when resistance to starvation and disease is higher. We call it baby beef, ton litters and cheap gains. It is correspondingly cheap health too.

Preventative measures have been disregarded but curatives were accepted. We have curtailed the consumption of the proteins and fresh foods under poor economics for health. We have taken to the hypodermic needle for health introduction into the bloodstream directly when health should be introduced into the body via the alimentary tract in the form of complete foods. Animals, too, are given limited proteins mainly as purchased supplements.

The problem ahead looms large now that we are internationally entangled, and with our food generosity taken for granted by the rest of the world. It is slowly dawning on us that soil fertility is a readily exhaustible resource. We may perhaps realize that more hospitals, more nurses, and more doctors are not the solution for failing health. Perhaps this realization will not come until half the crowd is in the rapidly multiplying hospital beds, and the other half of the crowd is trying to care for them. It is high time to learn that our national health lies in our soil and the guarantee against failing health lies in the wise management of the soil for production of nutritious foods. Fertile soils are the first requisites if we are to be well fed and to be healthy and thereby to remain a strong nation.

HEALTH DEPENDS ON SOIL

Our health depends on the soil. Bodies are really built from the ground up. In this time of crisis and thereafter all of us must build our lives and bodies from the ground up in more ways than one.

"In terms of the moon and the stars," inquired the Psalmist of old, "what is man that thou art mindful of him?" Man, the Bible tells us, is a mere handful of clay into which the Creator has blown his warm breath. Even in such early thinking some consideration was given to the possibility that the soil has something to do with the construction of the human body.

To the soil chemist, that which may be interpreted as merely allegory in scriptural language is in reality a great truth. Viewed in chemical terms, the adult human body of 150 pounds contains but about 5.5 pounds of ash, or the noncombustibles that came from the soil. This is the handful of clay into which all the processes of creation serve to blow the warm breath of sunshine, of water, of air and all else from above the soil, to build our bodies by way of the foods we eat.

Many are the failing bodies that reveal the weaknesses in their structures and in their functions. Few are the truly healthy bodies even among our draftees, from which rejections are more numerous than in World War I. Medical science is moving from cure to prevention. Many leading physicians point to poor bone structure commonly called rickets, and to disturbed body functions under the more understandable term of malnutrition, traceable back to the soil. Conservationists are joining with them and going even a bit farther back to fundamentals in thinking and searching for the causes of poor body structure in deficiencies in the very handful of clay from which we are made.

Man's body composition demands much calcium and phosphorus. Note in the following table of chemical contents of the body, of the vegetation, and of

Chemical Analysis of the Human Body in Comparison With That of Plants and of Soil

	Human Body %	Vegeta-tion % dry matter	Soil % dry matter
Combustile			
Oxygen	66.0	42.9	47.3
Carbon	17.5	44.3	.19
Hydrogen	10.2	6.1	.22
Nitrogen	2.4	1.62	
Calcium	1.6	.62	3.47
Phosphorus	.9	.56	.12
Potassium	.4	1.68	2.46
Ash			
Sodium	.3	.43	
Chlorine	.3	.22	.06
Sulfur	.2	.37	.12
Magnesium	.05	.38	2.24
Iron	.004	.04	4.50
Iodine		Trace	
Fluorine	Trace	Trace	.10
Silicon	Trace	0-3.00	27.74
Manganese	Trace	Trace	.08
Water	65		
Protein	15	10	
Carbohydrates		82	
Fat	14	3	
Salts	5	5	
Other	1		

See page 205 for a similar chart with a different breakdown.

the soil.

You will have observed that calcium or lime, for example, makes up 1.6% of normal body weight, yet appears in the available form in the soil to but .2%, though much more in total. Phosphorus, the companion element for bone construction, is in the soil usually in available form to an extent of less than .01%, an exceedingly small part of a total that in itself is of no great amount.

In the soil's capacity to provide these two nutrient elements liberally lies the solution to many problems in plant, animal and human nutrition. As the lime and the phosphorus are stintingly withheld or generously offered, there is exerted a quiet force that determines the ecological pattern for plants, for animals and for humans. In its quiet and subtle way, the supply of these two nutrients alone may nourish life forms into dominance, or may reduce them to annihilation. They are the first two nutrients in importance in terms of common deficiencies.

We dare not, however, focus attention wholly on but these two chemical elements, calcium and phosphorus, when fourteen of them are demanded for plant construction, and when sixteen, or two additional ones are required for animal body building. Four of these—carbon, oxygen, hydrogen and nitrogen—which constitute the bulk of all plant and animal forms, are not of soil origin but are supplied by air and water. Ten elements for plants, and twelve of them for animals, or three times the number of those of air and water origin, are drawn from the soil. The small quantity of each of them does not reduce the essentiality. Life is impossible with any one absent. It is thus in terms of absolute supply of body-building nutrients that the soil controls and directs the pattern of life. It limits life, at least, to those forms and numbers that can be maintained by the nutrient supply delivered by the soil.

The nutrient quality of soil depends on climate—not on the daily meteorological vagaries we call weather. But when these vagaries are studied over long periods of time and are given statistical treatment in large enough scope, to give us what we call climate, then we have the great force to which even the granitic mountains that are "rock-ribbed and ancient as the sun" must succumb, and by which they are changed to the "sluggish clod which the rude swain turns with his share and treads upon." In the climate at work on the rocks, are hidden the secrets of understanding how soils can be of variable nutrient value for life. We talk about weather but we fail to understand the climate. We are doing less about it. The climate is "doing" us into poor health as it moves the nutrients from the rocks to the sea and we make little effort to stem the exhaustion of the supply of nutrients in the soil.

Climate is further misunderstood when we had believed that plants are determined by the climate, instead of by the soil that was made by the particular climate. We have moved plants from place to place according to the plant's requirements of nutrition. Further, when one agricultural plant variety no longer is highly productive, we search and supplant it with another only to boast of our success in production of tonnage of herbage. But we forget that when the first variety failed because of the declining store of soil nutrients, then the second or introduced variety that succeeds must be producing tonnage by taking relatively less from the soil. It must be making itself by taking more from air, water and sunshine, or what is above the soil. Its service to animals must then be one of providing packing for empty paunches more than of supplying soil-contributed nutrients required for animal body construction. We are still viewing the ecological array of plants in terms of weather and climate rather than in terms of plant composition and the nutrients in the soil required to provide it. We need to understand our soils in terms of the climate producing them. We need to see life in all its forms related not directly to the climate, but indirectly to the climate as it determines the soil, and as the soil in turn determines how well we eat and therefore how well we live. Our health in truth depends on the soil.

Soils are developed to different degrees according to difference in climate. Rainwater is the chemical reagent to dissolve the rocks. Combined with carbon dioxide of the air, it makes the carbonic acid that leaves the acid residue of rock that we call soil, while much of the rock is dissolved and carried to the sea. A general rainfall map of the United States gives a basis for estimating the nutrient values of our soils. These soils are only partially developed to the west of the central or equilibrium belt, (20-30 inches of annual rainfall), and highly developed to the east and southeast of it.

Far west of this belt the soils carry soluble sodium and potassium, about all that has been weathered out of the rock. Lime is still mainly in the rock condition. Plant root feeding is more on the element in the rock form, rather than on those in the adsorbed form on the clay fraction of the soil. Plant feeding is therefore slow. Plant growth is scant both because of nutrient shortage, slow delivery to plants, and of rainfall shortage. We have, however, seen the scant rainfall as the only trouble, and have not thought of the nutrients as responsible.

Western soils in this moderate to scant rainfall have not been weathered enough to give much clay, or give the soil what may be called "body." They do not have much of this clay fraction in which the plant-soil activities take place. Unless the rocks have been extensively weathered, some of the less plentiful and less soluble elements will be deficient unless much rock has been weathered to concentrate enough phosphorus into the soil as residue to afford plants an ample supply. As a consequence of this under-development of the soil, cattle fed on western alfalfa hay as their roughage and western beet pulp as

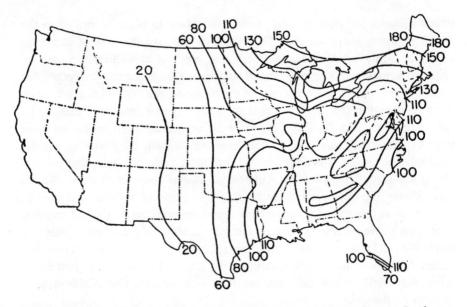

When precipitation is put against evaporation as ratios according to Professor Transeau, there are suggestions for the prairie conditions in the central states far east of the commonly considered prairie areas of low rainfalls.

their energy supply suffer a phosphorus deficiency disease known as "aphosphorosis." Given corn from the central cornbelt where its more weathered soils give higher phosphorus concentration in this seed as a phosphorus carrier, they do not suffer this health irregularity.

It is on these western soils that grass is the dominant vegetation. It accommodates itself to periodic shortages in rainfall by its own alternating periods of dormancy and active growth according to the moisture supply. Trees are uncommon on what we call prairie, but make their inroads along stream courses in the prairie areas. Even the tree species, so invading, are the cottonwoods, the Russian olive and others that can drop a share of their foliage as a means of accommodating their total leaf surface, or evaporating area, to the rate of water delivery by the roots from this soil. This grass vegetation of the plains of the West has delivered to us the nutrients from those soils in the form of beef and mutton for our consumption in the East. Short grass, highly concentrated of minerals and therefore rich in body-building capacities, has given us animals well developed in both skeleton and flesh and of unusual abilities and endurance.

Soils to the east of the hundredth meridian, on the other hand, have undergone much more rain and carbonic acid treatment. They have been more completely broken into other products. Much of the original rock body has been moved away. Many insolubles, like phosphorus, have become more concentrated in the soil. Because of the carbonic acid breaking down the rock, an acid clay residue results. The clay has taken on the hydrogen ion to give us the acid soils. When the clay takes on the hydrogen that has no nutrient value to plants, it gives up calcium, magnesium and other similar positively charged ions that are nutrients for plants. Acid soils are then deficient in calcium, for which limestone applied to the soil is a remedy. But we must not forget that soil acidity means the loss of other nutrients for which limestone is not the remedy.

Fortunately, while much water has been forming the soil, plants have been pushing their extensive roots through it to collect nutrients before they escaped. Organic matter formation in the virgin soils has been a tremendous force against nutrient loss by leaching. Continuous plant growth has been hoarding plant nutrients in dead plant parts. It has given us the fertile soils into which the pioneers put their sod-breaking plows. It was in terms of this reserve of soil fertility in the accumulated organic matter that the most eastern point of the prairies gave us the region of concentrated wealth. Unmindful of this fact, we have been mining this natural resource for shipment eastward, even to the Old World with little regard as to rate of its depletion, or as to significance for our own future national existence.

Should we move southeastward in going across eastern United States, so as to increase the temperature as well as the rainfall, then the soils not only contain more clay, but a different kind of clay. Its chemical nature is changed too. While the acid clay has a power to hold acid, these red, more weathered, or lateritic clays have little power of holding anything. They have little absorptive capacity. It is in these southern soils that there is no great acidity in the soil. Correspondingly, if the soils can't hold acidity, they hold no great store of nutrients. Here is the foundation in the soil, for some of the troubles commonly viewed as social or race problems when they are fundamentally in the nutrient supply of the soil. Rowntree in *Science* 94 (1941) 552-553 shows that seven out of ten men from Colorado were accepted as physically fit for military service but only three out of ten from one of the southern states.

Soils that are not so highly developed, nor extensively leached, have an ample supply of lime in the upper soil. As the soils are more permanently moist they lose their lime supply. Vegetation changes too with this change in nutrient store. Soils delivering lime liberally deliver less of potassium, for example, and are the native habitat for the nitrogen-fixing legumes like alfalfa. These mineral-rich forages are also rich in protein, and are a fit feed for bone-building and muscle-making of young animals. "We lime our soils for legumes" says the farmer, in acknowledging the fact that a liberal calcium supply must be put on the soils in eastern United States for the proteinaceous, mineral-rich leguminous crops. As we associate calcium in dominance in the soil growing proteinaceous forages, so we may associate potassium in dominance in the carbonaceous or woody forages, and in the forests. Putting it otherwise, since all forages are mainly carbonaceous, we may say that potassium is in dominance in forages which are low in proteins and low in minerals of nutritional significance. As the calcium decreases while the rainfall increases, or as the soil is more highly developed, there may be a decrease in potassium too, but as a much slower rate. Consequently, as soils are weathered, the ratio of potassium to calcium becomes wider and plants in nature shift from those dominantly proteinaceous and nutritionally mineral-rich to those that are deficient in proteins and minerals.

Under the principles by which the different kinds of vegetation are arranged we can readily understand the hardships endured by the Pilgrims on the eastern, wooded shores when they did well to shoot a few turkeys. It is no marvel that the "forty-niners" shot buffaloes by the score for their skins as they crossed the grassy plains of the West in the gold rush. Here in the vegetation in all its varieties are the colors and the forms by which the Creator has painted the picture of the nutrient qualities of the soil. On that canvas is laid out the pattern by which the higher life forms dependent on vegetation can guide their own distribution if they are to live healthy on the land.

Since "nature in the raw is seldom mild" we are inclined to give plants caveman tactics, in their getting nutrients from the soil. Quite to the contrary, plant life depends on what the clay fraction in the main consents to give it.

Plant nutrition is not a matter of hijacking the soil, but rather a case of trading hydrogen or acidity from the plant roots for the colloidally adsorbed elements that serve as plant nourishment. This is a process of colloidal contact exchange. The clay and humus take on the hydrogen or acidity from the root. They give calcium, magnesium, potassium and others in exchange. These are used by the plant, moved out of the soil-root region through plant synthesis and the way cleared for other nutrients to follow. Both the clay and the humus serve in this exchanging of the nutrients they had adsorbed on their surface from the soil solutions that contained them. The organic matter, or humus, may breakdown and deliver those more permanently held in the organic complex. Likewise the clay may break down, but even so its contribution of nutrients is less than that absorbed, and its offering by this means of no great significance. The clay and humus are the seats of activity in exchanging nutrients. They are the jobbers and it is on them in root contact, that the plant must depend for its nutrient store during the growing season. More humus, more clay, then, mean more nutrients and more rapid crop growth, provided these exchangers are not over-stocked with hydrogen, or acidity, the only item the plant can offer in exchange if it is to do any "growing" business. Plants offering hydrogen in trade to a soil already highly acid can do no more efficient business than the islanders, in the economist's illustration, who take in each other's washing for a living.

Clays are the plant's immediate source of nutrients, but will soon be exhausted unless replenished from the mineral reserve. Dr. E.R. Graham, of the University of Missouri, has recently demonstrated that the clay that becomes overstocked with hydrogen, or acidity, goes to the silt minerals and trades its hydrogen for nutrients and thus breaks down the rock fragments in the soil. Rocks and minerals are thus moving their nutrients slowly first to the acid clay and humus, and from there they are quickly taken by the plants. In this process there is the explanation of how nutrients are held in the soil against leaching by water, yet in large supplies quickly taken by the plant in the flush growth in the early season. Here we can see why clay soils are so much more productive than sands; why land that is turned out to rest grows better crops in consequence; why plowing as it rearranges the clay and silt can increase the supply of available nutrients on the clay for better growth; and many other aspects of better soil management reflected in more nutrient delivery by the soil.

In the nutrient quality of the soil, then, there is no magic by which any form of life is sustained. Rigorous natural laws rule life in any form. Only that life will long survive which obeys these fundamentals by which the complex chemical processes originating in the soil extend themselves to make life possible. Conversely, any form of life in its restricted locality is a reflection of the nutrient quality, or the fertility, of the soil. Conservation encourages our optimism as we understand the first source of the natural processes on which life depends and can thereby cooperate in, rather than hinder, the creation of more life and higher life.

OUR TEETH AND OUR SOILS

The knowledge about the human body and its many functions has been accumulating seemingly very slowly. The additions to our information have awaited the coming of each new science and the contributions by them in their respective fields. Dentistry as well as the medical profession has been ready and quick to accept and use any new knowledge that might alleviate human suffering.

In medicine, for example, one can list the major successive additions almost as separate sciences coming at the slow rate of about one per century. Anatomy was the beginning one, making its debut in the sixteenth century. The seventeenth century brought us physiology; the eighteenth added pathology; and the nineteenth emphasized bacteriology, all these for our better health.

Very probably the twentieth century will be credited with the addition of the science of nutrition as a major contribution to the better life of our people. Better nutrition is leading us to think less about medicine as cures and less about fighting microbes with drugs. In a more positive way, it is helping us to think more about helping the body defend itself by being well-fed and therefore healthy.

If we are to bring about good nutrition by means of good food, to build up a good defense for the body, that defense must be strong, not only against enemy invasions, as it can be against tuberculosis, but also against the degenerative diseases like heart trouble, cancer, diabetes. The science of the soil and its fertility, by which alone high quality foods can be provided, may well be an addition during the present century to our knowledge of the better functions and better health of our bodies.

It is proposed therefore in this discussion to lead you to think about the health condition of only one part of our body, namely, our teeth as they are related to the fertility of our soils.

In considering soil fertility as it provokes excessive carbohydrates but deficiencies of proteins and minerals, we need only to look at the chemical composition of the human body in comparison with that of plants (table 1). From these analytical data we see that potassium is taken into the plants in largest amounts of all the mineral elements from the soil, while calcium and phosphorus are next in that order. In the human body, these same elements are the major three, but calcium is first, phosphorus second, and potassium third.

Of amounts still higher than any of these in the human body is nitrogen; this is the key element distinguishing protein synthesized as amino acids from the elements only by plants.

Plants offer us mainly carbohydrates with only small amounts of proteins. Plant composition, considered as our food, represents possible shortages of proteins, of calcium, of phosphorus, and of probably other essential elements. We, like other animals, are constantly in danger of deficiencies of proteins and minerals, especially as we are more vegetarian.

By the very nature of the creative processes that start with the soil, carbohydrates are plentiful while there are deficiencies of minerals and proteins relative to the carbohydrates and fats. It is this nutritional need that encourage the carnivorousness and the use of animal products such as eggs and milk.

Soils naturally highly weathered are no longer well stocked with nutrient mineral reserves in their sand and silt fractions, nor with mineral fertility adsorbed on the clay. Such soils must of necessity give crops and foods which are mainly carbohydrates and are therefore deficient in proteins and minerals.

But quite the opposite, the less weathered soils under low annual rainfalls are mineral-rich in the silt and sand reserves and on the clay. Hence they give both proteins and minerals along with the carbohydrates in the plants grown on them.

In these facts we have the suggestion that any soil undergoing exhaustion of its fertility, whether by nature or by man, is bringing about a change in the chemical composition of any plant species growing on it. This change means that the plant species become more carbonaceous, less proteinaceous, and less mineral-rich. These changes occur within any single plant species, too commonly believed constant in its chemical composition regardless of the soil growing it.

BIO-ASSAYS RATHER THAN CHEMICAL ANALYSES

Many investigators continue to seek the functions of soil-borne nutrients," says Victor L. Sheldon,[1] "by correlating some morphological or physiological response with the quantity of some single inorganic element present in the plant ash. Nitrogen, as it is most often determined by burning the protoplasm in concentrated sulfuric acid, is no exception to this outmoded search. It remains a common practice, also, to multiply this total nitrogen value by some arbitrary factor, giving a dubious but readily accepted number, as the protein content of the material.

"In this study,[2,] it was often found that the suite of essential amino acids increased (often each one) while at the same time the Kjeldahl or total nitrogen in the dry matter decreased. This fact is shown in figure 1 for a non-legume.

"Arginine, threonine, isoleucine, valine, and leucine increased when phosphorus (increasing) was applied to the soil even when the total nitrogen (concentration) within the plants decreased. Conclusions based on the nitrogen content alone would have shown this a false criterion as to the status of the protein constituents (amino acids) in the plants.

"A similar situation is suggested for lespedeza (a legume). The addition of more phosphorus resulted in higher concentrations of valine, isoleucine, threonine and arginine while simultaneously the concentration of total nitrogen decreased from 3.6% to 3.2% (figure 2). Since it is the amino acids that are required in metabolism and not the elemental nitrogen *per se*, we need to adopt a newer criterion by which to measure plant production as it represents nutrition for the higher forms of life consuming it."

In his test applications of combinations of calcium, phosphorus, and potassium in contrast to their use singly in relation to the concentrations of amino acids in the forage crop, he says, "Wide variations would scarcely be expected when relatively small amounts of these inorganics were applied . . . Yet fully five years after these treatments were applied, the quality of bluegrass in terms of its constituent amino acids continued to be modified by the application of calcium, phosphorus, potassium and nitrogen . . . Where the applications of the first three elements were reduced and nitrogen added to balance the fertility better, the grass contained more of each of the amino acids.

"These data point directly to the need for balance among the elements coming into the plant from the soil."

In reading the preceding remarks which were hidden in a typewritten thesis, we are reminded of the higher concentrations of nitrogen that may occur in the greener forages when fertilized singly with that element by applications either through cow droppings or commercial fertilizer forms. But we are also warned that, while correcting the soil's deficiency of nitrogen as quickly reflected in exhibition of better color and more vegetative mass, that response is not proof

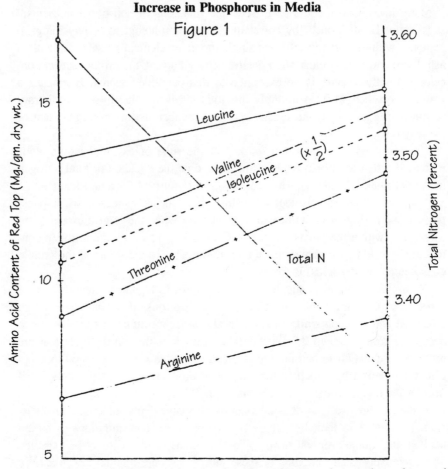

Increase in Phosphorus in Media

Figure 1

Concentrations of amino acids in redtop hay showing their independence of total nitrogen content when phosphorus was applied.

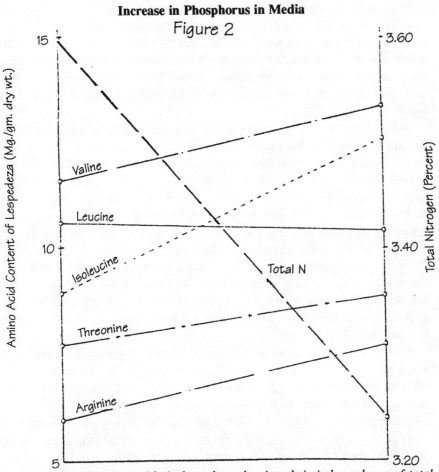

Increase in Phosphorus in Media

Figure 2

Concentration of amino acids in lespedeza showing their independence of total nitrogen content when phosphorus was applied.

of the nitrogen's making more protein, nor that the protein contains a more complete array of required amino acids. The latter result depends on the presence of nitrogen, of course, but that exhibitive element must be balanced by other elements operating, seemingly, more modestly with less showy effects waiting to be reported through bio-assays, that is, by grazing animals demonstrating healthier growth.

PASS-ALONG HUNGERS

Hunger is not a new problem. Next to sex instinct, it is the principal force driving man and beast into action. Today we understand hunger as worldwide in extent and importance. We are examining deeply enough to distinguish its "hidden" forms. We recognize these as due to shortages, not so much in the bulk of the food as in its nutritional qualities. We have not yet tagged all the different organic and inorganic compounds that provide these qualities, but have come to believe that many are grown into our foods. Consequently, we think about deficiencies in soil fertility as responsible for the failure of food to fully satisfy the body needs.

Hidden hungers are not experienced by man alone. Even the microbes within the soil have their hidden hungers. Organic matter of the soil, source of their energy food, accumulates in deficient soils and the microbes literally starve.

Sweet clover, fed as a green manure to the soil bacteria, may cause hidden hunger for potassium. While this popular, soil-improving legume grows and feeds ravenously on calcium, it can make bulk despite a meager supply of potassium. It grows well enough on a pile of crushed limestone suitable for fertilizer use. But, it has manufactured little potassium into itself and to satisfy the microbes, decaying it in the soil, should be supplemented by potassium which they must take from the soil. Thus, the corn crop, which is expected to benefit from this green manure as a supplier of nitrogen actually is robbed by it of potassium in the process. In such cases, the soil microbes, too, are struggling to cover their hidden hunger.

Whether these microforms of life may not be suffering hunger for other elements of fertility has not yet been fully established. We do not appreciate the fact that the wheat crop "eats at the second table," and that the microbes in seeking nitrogen literally pass on their hunger to the wheat crop.

It is important to note that both the corn crop and the soil microbes are well supplied with energy—one from the sun making photosynthetic compounds, the other from similar but decaying carbonaceous compounds.

Both, however, are suffering for small amounts mainly of nitrogen, by which their surplus energy foods can be converted into proteins and their diets properly balanced.

It is through difficultly-synthesized substances like the proteins that cell multiplication is possible, and by which the stream of life is kept flowing. Shortages of them really provoke the hidden hungers.

Wild animals well up in the biological scale have their hidden hungers too, though the fact is not always associated with the fertility of the soil.

Animals that are strictly herbivorous feeders are not commonly found on the highly leached soils of the tropics. Instead, buffaloes, elephants, antelopes and other grass-eating species are found on the prairies and savannas. They subsist on vegetation produced in areas of lesser rainfall, on calcareous soils, where natural legumes are abundant, and on soils which under cultivation produce the hard, or high-protein wheat. Soils given less to production of proteinaceous products and more to vegetation of carbonaceous contents give us forms of wildlife compelled to eat the seeds, the growing tips of branches, and other plant parts representing the maximum concentration of the plant's proteins.

FOOD QUALITY—AS PHYSIOLOGY DEMANDS IT

The old saying, "The proof of the pudding is in the eating thereof," is challenging the scientists to give us the values of what we eat in more specific characterizations than those of that simple—and usually pleasant—experience. Eating gives proof in terms of taste, but that is not yet specified by any standard, or reproducible, units of common agreement. We cannot report scientifically all that food does for us. We do not comprehend all the services of it, much less all the body functions in which it may play some vital roles.

Chemical analyses of the ash of vegetables of field crops are not highly informative. The very analysis destroys by ignition the organic compounds for which we eat food in the main. Vitamin assays, specifying these catalytic compounds, are more suggestive. Feeding tests, measuring the food quality by gains in body weight of animals, or by shifts in rates of body processes—of both animal and human—are more widely used to measure the nutritional qualities of food crops in relation to the inorganic fertility of the soil growing them. Though they are laborious procedures, assays of the rat, the Guinea pig, and other animals are about the only accurate measure of food quality now at our command.

Visible properties may be appealing, but they are not proof. The tomatoes serve as an illustration. The British market offers the housewife—and she seems to prefer it—a small tomato with fewer sections in it, with more acid and more juice. The American market has large, fleshy, many-sectioned, less juicy and highly flavored tomatoes. But what of the nutritional values, and what evidence thereof can we specify after the eating?

Food quality is arousing more concern in other countries under higher press of population than in ours. That may well emphasize nutrition and good health, as the reason for purchase of groceries for ourselves with decreasing tillable acreage of productive soil per person and declining fertility of those remaining areas under continued cultivation and higher costs. In Germany, and other European sections, a growing number of scientists, of magnitude sufficient for an international society, have concerned themselves in research in the nutritional qualities of different foods.

As an illustration of some of their work, there was reported a study of the qualities of apples used in feeding the children of two comparable orphanages. while the health of the children was studied and examinations of them made

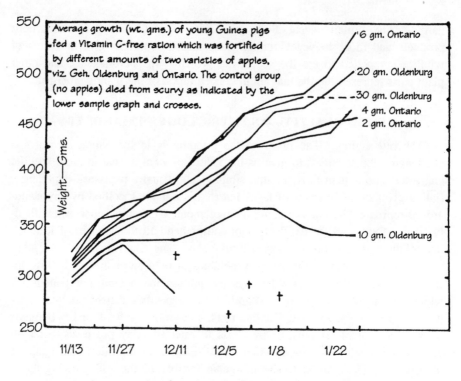

Average growth (wt. gms.) of young Guinea pigs fed a Vitamin C-free ration which was fortified by different amounts of two varieties of apples, viz. Geh. Oldenburg and Ontario. The control group (no apples) died from scurvy as indicated by the lower sample graph and crosses.

6 gm. Ontario
20 gm. Oldenburg
30 gm. Oldenburg
4 gm. Ontario
2 gm. Ontario
10 gm. Oldenburg

Weight—Gms.

11/13 11/27 12/11 12/5 1/8 1/22

Time Period—Dates

regularly, the apples were also put under bioassay by Guinea pigs for measure of vitamin C content.

Some of these results in summary (see accompanying chart) for a feeding test—from November through January—show that two, four, and six grams of one variety of apple, Ontario, were the equivalent in vitamin C content of 20 and 30 grams of another variety, Oldenburg, when used as supplements to a diet free of this essential nutrient.

Here proof is given to us in vital characterizations by the Guinea pigs. No apples meant their death; 10 grams of Oldenburg were but little more than survival; and as little as two, four, and six grams of the variety, Ontario, were the equivalent of from 5 to 10 times those weights of the former variety. If once we assay as accurately as this, not only the varieties of foods we grow, but each in relation to the inorganic fertility of the soil producing it, and if we specify the results as life and death for Guinea pigs, we shall no longer be indifferent to nutritional qualities of what we eat. Instead, we shall also be in position to improve and control nutritive values in what foods we grow.

NUTRITIONAL VALUES OF APPLES
ACCORDING TO CHEMICAL ANALYSIS

The apple has enjoyed great popularity, despite Biblical reports concerning it. Recently, however, it seems to have lost first place to the citrus fruits which are limited to more tropical climates and to areas of greater sunlight. We have learned that sunshine is a requisite when the vitamin C, even in vegetables, is synthesized more highly in relation to light than to the fertility marked by either the soil's organic compounds or any of its inorganic nutrient elements. Hence, variations in exposure to sunlight due to the shading of the fruit by leaves result in corresponding variations in the concentration of vitamin C.

Judging by this fact, it would seem that in the normally red apple, the higher degree of color development should indicate a higher vitamin C concentration. However, these two qualities are not linked, and there are plenty of red apples of "mouth-watering" appearance which are low in acid taste and in distinct and pleasing flavors. Also, there may have been so many apples grown with concern for no more than the red color and large yields that the choice apples, which might have held their markets on quality, have disappeared.

Considering vitamin C, let us remind ourselves that apples can produce this factor in the temperate zones and over enormous land areas from which the citrus fruits are excluded by shorter seasons of growth, lower temperatures, and less intense solar radiation. Accordingly, the apple needs to be maintained as a vitamin C source which lies within closer range of most persons than the tropics or non-frost areas.

The daily requirement of ascorbic acid (vitamin C) for adults in the United States is 50 milligrams. Using this as a basis, let us consider the nutritional value of three varieties of apples and three tropical fruits as shown on the accompanying chart. This shows that of two varieties a single large apple (larger than the citrus) will supply the daily adult need. However, of one variety of apples of smaller size, a half-dozen would be required to provide the minimum requirement of vitamin C. That requirement would demand consumption of 11 bananas. In terms of these facts, the apple must be considered a good source of ascorbic acid, especially in view of its availability over long seasons, its extended storage life with slow loss of vitamin quality, its small amount of inedible waste, plus other properties surpassing those of citrus fruits.

In Dr. Werner Schuphan's studies,[1] 134 different kinds of apples were measured for vitamin C content, even though this factor is not considered in judging market value. The amounts of vitamin C range from a high of 31.8 milligrams in 100 grams of fresh matter to a low of 2.3 milligrams. Apples of 16 kinds recommended widely in Germany include those above 14 milligrams. Since the inedible portion of the apple is small it requires not more than three

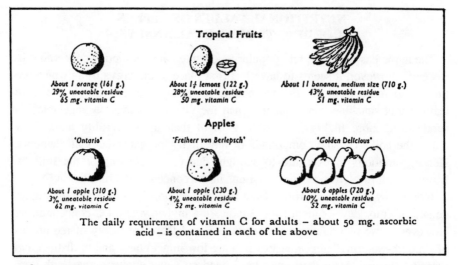

Tropical Fruits

About 1 orange (161 g.)
29% uneatable residue
65 mg. vitamin C

About 1½ lemons (122 g.)
28% uneatable residue
50 mg. vitamin C

About 11 bananas, medium size (710 g.)
43% uneatable residue
51 mg. vitamin C

Apples

'Ontario'

'Freiherr von Berlepsch'

'Golden Delicious'

About 1 apple (310 g.)
3% uneatable residue
62 mg. vitamin C

About 1 apple (230 g.)
4% uneatable residue
52 mg. vitamin C

About 6 apples (720 g.)
10% uneatable residue
52 mg. vitamin C

The daily requirement of vitamin C for adults — about 50 mg. ascorbic
acid — is contained in each of the above

or four medium or small apples per day to supply the adult needs of vitamin C.
If we were to take 30 milligrams of C, which is considered the adult requirement in England, one or two apples a day would suffice. The 16 different kinds
(with some periods of storage) provide apples most of the year around, except-
ing the four months of summer.

Included in 134 different kinds of apples tested, and listed (without reference
to locality of growth) in their order of decreasing contents of vitamin C, are 10
familiar American and English varieties having more than 15 milligrams of
vitamin C per 100 grams. In the order of decreasing amounts of C, and the
numbered order of rank, they are as follows: 2—Pippin (Ribston); 6—Golden
Noble; 4—Ontario; 16—Yellow Belleflower (Linneous Pippin); 17—King of
the Pippins; 20—Lady Hollendale; 22—Northern Spy; 24—Norfolk Pippin;
28—Bramley's Seedling; and 32—White Transparent.

Without the numbers of still lower rank (from 33 to 84), the list included 10
additional familiar names of American and English apples. Accordingly, the
apples of both these countries must be expected to vary in vitamin C from a
high of 30 milligrams per 100 grams of edible fresh substance to a low of near
zero. One must, therefore, learn about apple quality since some of this fruit can
deliver such high amounts of this one nutritional factor of which we require
less than a 50th of an ounce per day.

In contrast to citrus fruits, apples do not contain harmful oxalic acid. We
cannot boast about either the apple or the citrus as sources of quantities of
mineral substances, including the trace elements. "Apples contain potassium,
magnesium, iron chloride and phosphate in much smaller quantities than are
present in many vegetables which are considered low in these elements. Also,
the sulfur and sodium content is no higher than in the vegetables having the

smallest contents.

Vegetables contain, for the most part, multiples of the values of apples for these eight mineral substances, as well as for trace elements. Here are a few examples: Calcium—Kohlrabi contains over 50 times as much as apples; Phosphate—Cucumbers contain more than 25 times as much; legumes, up to 100 times as much. In this connection, it is of interest that oranges, like apples, fall far below the vegetables insofar as a generous supply of mineral constituents is concerned."[1] The fruit at the end of the limb on a tree is too far removed from the minerals in the soil to compete with the vegetables which are closer to the mineral source.

In the content of pectic substances (jell-makers) and of tannic and malic acids, which are substances of significant dietetic and therapeutic values, the apple has a decided advantage over the orange. The latter has only the refreshing taste of citric acid. In addition, the orange is reported to contain 24 milligrams of oxalic acid per 100 grams of fresh substance.

A recent scientific report cited the unique ability of malic apple acid to serve as a "chelator" of the mineral elements, mobilizing their activity from the soil of the plant. Iron, put into a large, synthetic, organic-chelator compound (ethylene diamine tetra acetic acid, or EDTA) applied to soybean roots was soon found in the exudate from the cut-off stem. But instead of the iron's being still chelated by the applied EDTA, it was carried within the plant in chelation, mainly by malic acid, but also by malonic acid, both of which are by no means of as complex chemical structure as EDTA.[2]

Here is a suggestion of nature that should prompt some biochemical research on the biological behavior of the inorganic and dietetic assets of the apple. Possibly we might learn the physiological bases for the widespread empirical belief of many years that "An apple a day keeps the doctor away."

NUTRITIONAL VALUES OF
VEGETABLES CHANGE RAPIDLY

In these current reports on the quality of food crops (vegetables and fruits), the Bloomsdale Savoy spinach was the first one considered. This is sometimes eaten as a fresh vegetable (plant tops only) in a green salad, but is more commonly taken cooked as a potherb. It may serve here in considering the quality of plant tops; carrots will serve for quality of plant roots; and apples for quality of tree fruit. These three have a long history of widespread acceptance as food. By means of these, through chemical and biochemical measure—if not through biological assays where quality as nutrition can be measured accurately by the growth response of microbes, animals and man.

The fact that spinach is grown almost all year around creates varied quality, particularly from the biochemical standpoint. Plants are living substances, anchored to the soil in place and fed according to the soil's biodynamic processes which furnish nutrition to the plants. Simultaneously, the soil microbes may be collaborators or competitors eating at the first sitting.

Different soils and different weather conditions (acting on both soil and quality may become more quantitative plant) at various times of the year must be expected to bring about changes in the quantity and quality of the constituents created within the vegetable crop. Spinach gives opportunity to consider changes in nutritional values during the growth, or life, of the plant, and then after the harvest, or death, while en route from the garden to market and the consumer's table.

Some experiments with spinach[1] report these chemically-measured changes from the time of seeding through growth, divided into five stages or crops. Seeding was done in mid-January in forcing frames, and a light fertilization was given at the beginning of March. In the middle of that month, the first plant sampling was made. This consisted of plantlets which had developed rounded and transitional leaves in addition to the cotyledons. The second sampling, at the end of March, included additional small, successive leaves. By the third sampling, these leaves were better developed for harvesting as young spinach. The fourth sampling, or crop, gave plants for customary early harvesting, while the fifth sampling, April 20, 95 days from seeding, consisted of fully-grown plants like those one finds on the market in spring.

From the chemical studies of these five crops, or samplings, during the plant's life, the following information was reported:

1. The undesirable oxalic acid content diminished with progressive development of the plants.

2. The total amount of sugar (including both monosaccharides and disaccharides) is seemingly adapted to the metabolic state of the plant. It was highest at the third sampling.

3. Ascorbic acid concentration (vitamin C) rose to its maximum in the second and third samplings, and then fell to its lowest at the time of full growth, when the plants are commonly marketed.

4. The concentration of carotene was highest in the youngest plants sampled. It was lowest in the last sampling, being about half of the initial high. The increase in leafstalk and in the mid-rib of the leaves, both of which contain no carotene, makes the increase in plant bulk serve to dilute the concentration in the plant as a whole.

5. Of the essential amino acids, some were high during the second sampling, and most of them declined from then on, or from the first sampling. Accordingly, the eight essential amino acids, commonly called "the index of biological

value," were at a maximum during the second of these samplings, and declined from that early degree of plant development to their lowest level at the time of harvest for market.

6. The nitrate nitrogen increased significantly in the last two samplings. With an earlier dressing of nitrogen, phosphorus and potassium on the soil at the beginning of March and another one of nitrogen only, the time period from then until late April should permit enough nitrate production in the soil to account for the high nitrate content in the crop at harvest.

7. The inorganic elements in the crop (excluding sodium) did not show significant changes for the different stages of plant growth. Sodium and chlorine increased in concentration toward maturity of the spinach plants.

8. Sulfur behaved much the same, with but a slight increase toward maturity, as was also true for calcium. But the latter was more pronouncedly so, in a kind of inverse effect to the concentration of the oxalate. While the shift for oxalate was from a high to low in a ratio of 4.7 to 1, the shift by calcium, as inverse from low to high, was in a ratio of 1 to 4.7. The harmful effects by the oxalic acid are reduced through its precipitation by calcium, but the beneficial effects of the calcium are thereby eliminated, especially in the spring. The crop in autumn was lower in the oxalate.

After harvest, with the plant top cut from the rootstalk and sustenance from there eliminated, the metabolic processes will be under decided stimulation of the enzymatic actions. This reaction doubles its rate for about every 18° F. rise in temperature at which the harvested crop is held. Consequently, if cold storage is not used, storage loss at 65° F. For 40 hours may amount to nearly 20% by weight, according to Schuphan's data. The protein nitrogen, as a percent of the total nitrogen, may diminish by nearly 12%.

This latter change alone, when the living tissue is consuming itself, stands out as a loss which is not measured by the same scales as those used to measure moisture and plant bulk. The former loss is of interest to the nutritional scientist; the latter to the dealer and the economist, both of whom are far removed from consideration of the plant's waste of its own proteins, sugars and other quality factors for better health and taste.

Unfortunately, the change in quality of vegetable plants during life and after harvest are recognized mainly in terms of their "filling" power, rather than in terms of nutritional value.

There is much yet to be learned concerning the nutritional qualities of food plants.

THE LITTLE THINGS COUNT IN NUTRITION

We were startled a few years ago to discover that a little of a trace element, like manganese, missing in a chicken's ration would let the chicken go down on its hock joints, or its elbows, merely because the bones failed to develop the little stirrup-like growth through which a tendon must go and be held from slipping off.

What has manganese got to do with that trouble when we analyze the bone chemically and find little or no manganese there? It illustrates deficiency in our thinking that needs notice. We fail to realize that ash analyses are no complete index as to what we must provide by the way of the soil to make body construction complete. You can analyze the bone and find that the manganese isn't there in any significant amounts.

But the manganese was there and served as a tool somewhere along the line. Perhaps it shaped something or synthesized something that must come in to play if that chicken is to keep from going down with those slipped tendons. We must remember, that in these syntheses of all it takes to make a healthy body, many of these things that come from the soil are tools along the line and not products of the final construction.

You could analyze this building today and you would not find a single mason's hammer or trowel in it. Would that prove that the trowels were absent in all of the history of this building and that none are essential for buildings? They were here at a certain time, but now they are gone.

The chemical analysis of this building will not give you the details by which you could construct another one in good order. Then, too, you wouldn't need 10 times as many trowels to build a building ten times as big as this one. The same number of trowels that were in this building during construction will build a building ten times as large, but it will take a little longer time.

Time is what nature takes to build or to break a body. So this matter of making a chemical analysis of a body and then interpreting it as a guide to build another one is entirely fallacious and insufficient. Yet many health practices are founded on conclusions no more logical.

By that token, we can talk about the trace elements. We don't know where they are in the body or where they function in all cases.

You don't think of trowels unless you see a building being built. You can analyze all of the buildings in the world and it will never dawn on you from that approach that you use trowels or levels or chalk lines or scaffolding. And it is in that same viewpoint—that there is much to be considered as coming from the soil as tools—that up to this moment we haven't considered as essential.

Of course we haven't really seen nature perform. Body growth isn't a motion picture flashing on the screen. It is nature's secret in many phases yet. If you started out to build a building and didn't have some trowels, you might have some deficiencies in that building. You might lay those rocks up some way or other without those tools, but sometime later the rocks would shake apart, and the building would fall and crash because there were defects that came in during the process of building that structure.

So in the process of growth (quite more complicated than the process of hanging on fat), there are times when the deficiencies as small as those of the trace elements can become significant. Let's remind ourselves then, that in this problem of providing material for construction and the tools for erection of a body if we should have a shortage in any of those performances, or a low quality somewhere along the line, we must expect the final product to suffer accordingly.

Too often when we have a manifestation of ill-health, we say that it is "bad luck." If, for example, you have a calf that is born in August, and the season goes into a droughty autumn so that the mother cow doesn't have very much to feed on, the cow can not give much to the calf. If it was a droughty summer while the calf was being nourished as a foetus, and both cow and calf go into winter dry feed until spring comes along, it ought not to be startling or surprising if in April that calf is down with some trouble suggestive that the femur bones had snapped at the hip joint.

We say, "We had bad luck—those calves did not go through the winter." Of course they didn't. What is the cause of the trouble? Is it the winter? Is it the cold weather? No, it is the poor nutritional level which that season has provided.

NUTRITIONAL QUALITY OF VEGETABLES
VIA PLANT SPECIES AND SOIL FERTILITY

When we speak of nutritional values of any of the vegetable crops, do we have a clear concept of what part of the plant represents those values, and how widely they may vary? They will vary naturally because of different plant species within a single use category (for example, "greens") and within the same species, because of where, when and how they are grown in your own garden. They will vary on the market, also, because of from where, when and how the crop was delivered. Also, they will vary between the time of purchase and time of consumption.

Vegetables grown under carefully controlled conditions and put under chemical and biochemical assay for such values (reported in Werner Schuphan's book[1]) serve to shock one's faith in the pedigree of a vegetable crop as any guarantee of its health-building value when eaten. The pedigree is merely a mark of the potential, or the possibility of values. Only the plant's nourishment from soil supporting the potential of high nutritional quality can assure this latter property.

The biochemical understanding of the life processes has been carried back as far as the two bits of protein, namely, two nucleic acids, of the "gene" of the cell. One of these acids serves as the warp and the other as the weft in weaving the structural pattern. The infinitude of these two nucleic acids eliminates any extensive power for carrying very much control of growth from the parents to the single mated cell. The many growth stages, pushed along by many growth factors from conception to full development, offer numerous chances for deficiencies and disruptions of the potential in nutritional values of the plant products for higher life forms.

In the case of the vegetable "greens" for human consumption, a variety of different kinds of plants is included. Some of these contain considerable oxalic acid, a byproduct of the plant's metabolism, while others do not. Oxalic acid reacts with the nutrient element calcium to form insoluble and indigestible fine needle crystals. These are irritating to the tongue and disturbing to the taste.

While green, leafy vegetables may appear to be a rich source of dietary calcium, according to analyses of ash contents some species would not classify as such under tests of digestion and nutrition, while others similarly listed as "greens" would. Savoy spinach, Swiss chard, beet greens and New Zealand spinach, all belonging to the goosefoot family (Chenopodiaceae), may be worthless as contributors of calcium and magnesium. In addition, if cooked with milk, their high oxalic acid content may bring about unavailability not only for their own calcium and magnesium, but also for a large percentage of these factors coming from the milk or other accompanying foods.

However, such is not the chemical composition, digestive behavior and nutritional quality for these mineral nutrients of the "greens" of the mustard family (Cruciferae), including kale, turnip greens, collards and mustard greens—to name four. All of the eight species of vegetable plants listed offer considerable calcium. But, considered nutritionally with respect to their contribution of calcium (the most necessary of the mineral essentials) the two families differ sufficiently to warrant more careful discrimination in their use (at least in children's diets) than is commonly exercised.

Studies at the Missouri Experiment Station[2] showed that the oxalic acid content of Savoy spinach, and its relatives, is seriously disturbing; but the "greens" in the mustard family are practically free of oxalic acid. Also, very often the latter are higher in vitamin content. Since the plants of the mustard family offer so much more mineral value than do those of the goosefoot family, it would be worthwhile to learn the nutritional values of the various "greens" so one can make an intelligent selection.

The Missouri studies demonstrated not only differences in oxalic acid content according to plant species, but also showed these differences in relation to varied additions of nitrogen and calcium as soil fertilizer. (See accompanying graph.) Chemical data for calcium, magnesium and oxalic acid showed enough of the latter to render both the calcium and magnesium insoluble; and in some cases showed nearly double the amount required to make both calcium and magnesium unavailable. The oxalic acid content was modified decidedly in relation to each of the fertilizing elements, calcium and nitrogen.

Here is clear evidence that the nutritional quality of vegetables may depend on the fertility of the soil. These studies show that variations in the plants' contents of the cations calcium and magnesium, as well as of the anions nitrogen and phosphorus, resulted in much greater variations in the production of oxalic acid when only two nutrient factors of soil fertility (nitrogen and calcium) were variables in the soil treatments. Your health depends on the soil that nourishes it.

VARIABLE QUALITY PRODUCTION BY FOOD PLANTS

From the standpoint of economics, we emphasize the quantity of vegetables

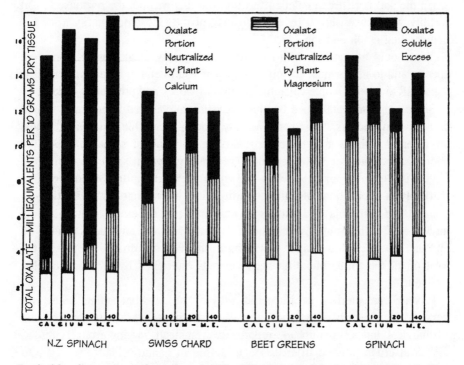

Probable dispostion of oxalate in New Zealand spinach, Swiss chard, beet greens, and spinach when grown at variable levels of calcium.

and other foods available on the market. From the health standpoint, we must look to their quality as nutrition for a biochemically complex, warm-blooded body. There are accurate measures of quantity, on which seller and buyer are well agreed. However, the quality, much unknown to both, is seldom critically considered prior to purchase and consumption. Too often it is assumed that quality parallels plant pedigree. In view of these facts, it is fitting to consider just what factors in food crop production determine quality, and how these factors may take different orders of significance for different crops.

In the previous section, eight crops of vegetable "greens" were classified according to how they synthesized the metabolic byproduct oxalic acid, in the presence of which calcium and magnesium react to form calcium and magnesium oxalates. These are highly insoluble, hence indigestible.

It is significant to note that Swiss chard, Savoy spinach, beet greens and New Zealand spinach—the "greens" of the Goosefoot family which produce the oxalic acid—are also low in calcium (less than one percent of the dry weight); while those of the mustard family, which do not produce oxalic acid, are twice as high in calcium. They are also higher in their concentration of that element when the soil is richer in calcium fertility. But this was not the response to soil fertility which was shown by the "greens" of the Goosefoot family. (See accompanying graph on page 245.)

You should be familiar with the nutritional quality innate to the Goosefoot species and to its response to soil treatment if you intend to feed children these "greens" to furnish ample calcium for building bones and preventing rickets.

Some crops are catalogued by characteristics emphasized in common garden conversation, such as, "Oh, that is an easy crop to grow. One can raise it almost anywhere." This is a statement commonly made about spinach, either the Bloomsdale Savoy or New Zealand. These are grown in spring and autumn. Savoy is higher in oxalate when grown in the spring. Both kinds respond markedly to applications of nitrogen fertilization, but tolerate wide variations in contents of minerals and protein. They are, consequently, grown almost anywhere. But their nutritional values also classify almost anywhere "on the up and the down," with the latter prevailing in the case of "crude" protein as the total of all the forms of its nitrogen contents multiplied by 6.25 and labeled "protein."

Protein, characterized as a specific array of the essential amino acids, is a very good measure of nutritional values, or of what is often called the "Biological Index." Schuphan[1] reports fresh spinach varying from 1.38% to 5.38% in crude protein; from 0.31% to 1.46% in oxalic acid; from 1.10% to 2.75% in total sugars; from 0.05% to 0.25% in total calcium; and from 0.02% to 0.12% in total phosphorus. All of these are variations from the low to a high of five times the former, or 500%.

When the composition of a plant can play over such a wide range of its synthesized contents and of essential nutrient elements, the soil fertility does not appear to be a serious limiting factor. In that respect, we can understand why spinach can be grown on almost any soil; as a field crop in spring, or in autumn; and as an over-wintering vegetable in cold frames or in hothouses. Such range in possibilities for growing its bulk ought to raise the question, "What is that bulk in terms of quality as protein which goes as low as one percent in 'crude' protein?"

Nevertheless, before we condemn this "greens" vegetable, it is significant to note that it is not completely devoid of any of the 10 essential amino acids with make up complete protein. This was determined by extensive samplings in the studies by Dr. Schuphan. Also, using vitamin C (ascorbic acid) as another

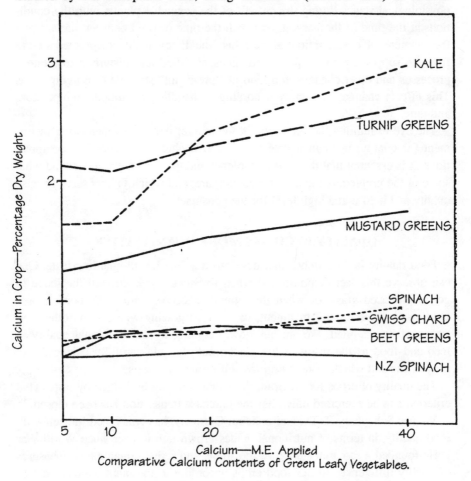

Comparative Calcium Contents of Green Leafy Vegetables.

criterion of nutritional value—along with the mineral elements and the complete proteins—this vegetable does not grade at such a low level. In nearly 300 samples of spinach, Schuphan reports the variation in vitamin C from 2.4 milligrams per 100 grams (24 ppm) to 157 milligrams per 100 grams (1,570 ppm), or from the low to a high of 65 times greater.

In carotene, the precursor of the fat-soluble vitamin A, spinach varies from 0.7 milligrams per 100 grams (7 ppm) to 6.3 milligrams per 100 grams (63 ppm) or from a low to a high of nine times greater. When spinach can reach those two high values in ascorbic acid and in carotene, while red, not quite ripe, tomatoes carry respectively 41.7 and 0.59 milligrams per 100 grams (417 and 5.9 ppm), the spinach does not take a low rank in these essentials in quality as food.

It is important to realize that food constituents vary with the respective physiological stages during the growth of the vegetable plants; with the conditions at the time of the harvest; and with the time elapsed between harvest and the moment of consumption, since when the living plant is uprooted or cut from its supporting rootstalk, it undergoes modified and disturbed respiratory processes and other changes in a kind of "pre-death" struggle of a living body. This effects changes in quality according to handling conditions, temperature and time.

The eye examining the food crop at the market is not a fit measuring instrument for quality; nor can it determine internal changes in chemical compositions. It is essential that the market gardener and the customers understand what some of the problems of quality production are, and which factors serve to raise quality and hold it at a high level for the consumer.

QUALITY BECOMES MORE QUANTITATIVE

Food quality is the essence that develops a noble, health-guarding taste. One can observe this fact demonstrated when livestock break through the fence to get to preferred grass, or when the animals choose to graze the portion of a field given a certain soil treatment, or to avoid grazing where some other soil treatment was applied. We are but slowly coming to see nutritional quality in feed and food as the result of the plant's nutrition via carefully balanced fertility of the soil which insures healthy, self-protecting crops.

The quality of crops for, or from, the market can not be judged by taste. That criterion can be exercised only after the purchase transaction has been closed.

Werner Schuphan, Ph.D., Geisenheim, Germany, has concerned himself with food quality, in terms of nutritional values grown into it, ever since World War I. He founded a research station of the study, and refined quantitative measures of quality differences in our food crops according to the many modifying factors. But it was only this year that his assembled reports appeared in the English translation. These comprise the 280-page book, *Nutritional Values in Crops and Plants (Problems for Producers and Consumers)*, published by Faber and Faber, 24 Russell Square, London, W.C.I.

According to the author himself, "The book defines a position with regard to worldwide problems of the quality of agricultural and horticultural food plants. It attempts to integrate concepts of all those interested in the problems of quality in order to find a common denominator."

"The demand for quality—this must be emphasized—is not confined to the consumer," he continues. "Both producer and merchant are interested for similar reasons. This does not mean they have the same reasons regarding what is meant by the term 'quality.' Moreover, the interests of those concerned in the quality problem are different. Producer, manufacturer and retailer often attend to their own economic interests too one-sidedly. To direct these largely to the legitimate desires of the consumers for an unobjectionable, wholesome diet is a far-sighted policy which is also an advantage to others who are interested in plant reproduction.

"On the other hand, the consumer must not make unrealizable demands on our field and garden products; efficiency in production in agriculture—as in industry—is the main thing. Our modern methods of cultivation are subservient to this necessity. With the daily increase in the world's population, rational production is the only way to provide quantities of food."

"However, bulk production of attractive crops is not always synonymous with high nutritional quality," Dr. Schuphan explains. "Therefore, constructive criticism of the methods employed by the producer, the manufacturer or the retailer is appropriate where loss of quality touches the interests of the consumer.

"These questions deserve to be given particularly careful attention by those responsible for the nutrition of infants, of children and of the sick."

Dr. Schuphan's book outlines the problem and considers quality as the measure of value in terms of, first, the consumer's expectations under commercial grading in various countries; and, second, the mounting attention to utility and physiologically nutritional values. These emphasize what is called (a) "biological value" as an expression of quality, (b) protein supply, and (c) taste.

Nearly half the book is devoted to clearing up the confusion about how much and when the hereditary assets of the plant species can contribute to its nutritional quality; and, similarly, how the environment, natural or man-managed, can be a factor in that respect. It is here that the many tests carried out at the research station under the leadership of Dr. Schuphan are presented as quantitative measurements of nutritional values in their several phases as we now appreciate them.

Differences in nutritional values because of species, varieties, "cultivators,"[1] form and shape, climate, soil, cultural practices, spacing, manuring, chemical protectives, time of harvest, transport, size, weight, storage, preparation for market, and other factors are presented by specific test data to make this volume an extensive—if not complete in all aspects—reference volume on quality of food as nutrition.

Food quality as a therapeutic factor is discussed also, particularly as a preventive for such degenerations as heart disease.

"In this book," says the author, "the interests of the producer are to be brought into harmony with the desires of the consumer. This involves confidence. Confidence presupposes knowledge of the problem. It is the object of this book to provide the knowledge."

SIZE AND WEIGHT VERSUS NUTRITIONAL QUALITY

We have accepted size and weight as quantitative criteria for the market qualities of most of our agricultural products. This considers their services and values as foods mainly in terms of satisfaction. That is expectable in the struggle against hunger for which the first concern is to get enough bulk or weight of food to eat and satisfy the craving by hunger.

In the business of agriculture, we are aiming for high yields per unit area of soil surface, or per acre. The first thought for years of bringing that about was the hope for larger units of the harvest. That emphasized growing bigger ears of corn; larger fruits like apples, peaches, and pears; larger carrots, turnips, sweet potatoes, Irish potatoes and other tuberous or root crops; and also bigger heads of cabbages, lettuce, cauliflower; whether the seed, the root or the leaf of the plant was the harvested edible portion. For many years the selection, and propagation for enlarging the harvested unit have been the major struggle for increased crop yield.

That had its economic advantage. It brought a larger edible unit with lesser waste as the inedible portion in preparation for preservation and consumption. It is a simple mathematical fact, that according as a near-spherical unit is larger, the less surface there is per unit volume or unit mass of contents.

Public exhibition of agricultural products, as invitations for purchase, have also invited more size and weight; prize-winning products catered to those characteristics. But now with increase in mechanical over manual preparation for consumption, like peeling potatoes, fruits or other food items, the larger size is not considered a significant factor.

In recent years the yields per acre have been pushed up by aiming at smaller units but more of them by thicker plantings per row and less distance between rows or more of those per acre. As a consequence, the quality as nutrition is now coming in for closer scrutiny. With recognized decline in the percentage of crude protein in our corn grain from 10.30 to 5.14%, as illustrated by that grown for exhibition[1] in corn husking contests, and with accumulations in the plant of unmetabolized, or nitrate, nitrogen and its hazards as poison, the question is widely considered whether the nutritional values per smaller units cannot be lifted significantly.

In the first place, we need to remember that size and weight are genetically controlled. Variations in those are provoked by ecological influences, but though they occur their limits are set by birthrights. Temperature, water supply, and available nutrients in the soil are effective in giving more unit size and weight. It is common experience that generous supplies of water and nitrogen give bigger potatoes, cabbages, apples and other crop units. But those as external and physiological factors are coupled with the genetic or internal factors. The latter we can only choose and accept. The former we aim to manage and to improve.

In spite of the more ready acceptance of the genetic matter of size and weight limits, even they can be modified to our advantage biochemically, nutritionally, storage qualities and otherwise, by conscious effort in selection of the specimens, therefore, in the course of plant propagation.

For the relation of size to nutritional quality let us consider seed first and look at the pea. For this vegetable, it is almost a common contention that peas are more tasty, especially sweeter and less tough, according as they are smaller. That, however, is apt to be due to less of maturity and an early stage in the ripening or moving towards the end of physiological activities and dormancy. Should we not expect a certain limited stage in that process to represent the maximum in the several items making up the most pleasing taste and collection of nutritional values?

The researches by Dr. Schuphan[2] report the concentrations of sugar and of vitamin C going lower with increasing physiological ripeness of the pea-grain. The pure protein behaves in the converse. That rises very suddenly after the pea-grain size reaches 8.2 mm. The tenderness and the taste quality diminish then.

Dr. Schuphan has made analyses of peas of different sizes to note the composition of those in terms of their essential amino acids. Data for one variety of peas are quoted in table 1. Of the ten amino acids, for five of them the concentration increases with size or maturity; for two of them there are decreases; and for three of them they remain nearly constant regardless of size or maturity of the pea-grain.

Since the seed is the power of maintaining the species in successive generations, one is dealing with genetic values and would scarcely believe size a significant factor save as stored reserve energy foods since the germ is very small and we cannot expect to measure its qualities by simple inspection. Quality in that respect is still deeply hidden. Studies of nutritional qualities of forages for livestock have very lately been telling us that size and weight, or bulk, per acre are deceptive as to nutritional values. Livestock choices guiding managed grazing tell us that the younger growing plant, not the large yielder or mature one, is the most concentrated and more nearly balanced in nutritional values for growing warm-blooded bodies. It is high time that more attention be given to nutritional values of vegetable products along with attention to economic values of bulk production.

A NEW BOOK—"OUR DAILY POISON"

This book of 178 pages, written by a scientist with agricultural experience, is a forceful presentation of the facts which a few advanced students have recognized in medicine, health, biology, ecology and soil in their relation to the present increasing use of deadly poisons as a hopeful solution of the complex biological problems of agriculture, food and public health.

The discussion of the subject is divided into 11 short chapters. Two of them discuss the poisons used as sprays for macro- and microscopic pests of plants and animals; three chapters are given to fluorides now on trial; one is on the new cosmetics, with emphasis on the skin as an absorber of poisons; one discusses the perils of hormones; one explains the dangers from our use of vaporizers and fumigators; one reports the poisoned soil and plant absorption of poisons resulting from these technological weapons used against microbial life above the soil while disregarding their dangers to the beneficial soil microbes; one chapter is given to our doctored daily bread; and a final one outlines the limited choices facing us under spraying practices.

The evidence cited is taken from the several voluminous Congressional reports of the Hearings before the House Select Committee to Investigate the Use of Chemicals in Food Products. These hearings cited extended from September, 1950 through March 1952.

Mr. Wickenden brings these extensive reports by many scientific and industrialists into simple, condensed presentations of the major conclusions which are apt to remain buried in the Congressional files and unappreciated as the reasons for some of the laws resulting from those hearings.

The discussion of this subject indicates that our viewpoint about health is becoming broader. We are coming to see health as the result of a nutrition which grows a self-protecting body according to the same biochemical laws which were operating during the evolution of microbes, plants, animals and man—even before man assumed that he could exercise control over the entire biotic pyramid by means of chemicals of his own compounding.

TABLE 1

Composition of Green Peas (Calcium as Percentage, Essential Amino Acids as Gm. per 100 Gm. Crude Protein)

	"Kelvedon Wonder" 1956		
	Small	Medium	Large
Dry Matter	6mm	6-8mm	8-10mm
% Fresh Wt.	18.40	19.71	21.55
Calcium %			
Fresh Wt.	5.63	5.44	5.56
Dry Wt.	30.6	27.6	25.8
Protein Nitrogen	43.4	44.9	60.8
% Total Nitrogen			
Amino Acids, gm.			
Valine	4.4	41	4.0
Leucine	3.8	3.8	5.4
Isoleucine	3.6	3.9	4.6
Threonine	5.1	4.0	3.5
Arginine	7.7	12.2	14.3
Histidine	1.7	1.5	1.7
Lysine	3.5	3.7	5.1
Phenylalanine	2.2	2.3	2.9
Tryptophane	0.7	0.7	0.7
Methionine	0.7	0.8	0.6
Total	**33.4**	**37.0**	**42.8**

It exhibits the growing appreciation, among some of the medical profession, of the broad ecological basis of health of ourselves, when that basis is now being recognized under epidemics by some concerned with public health. Some of those officials are no longer satisfied with finding the microbe and its carrier, in an epidemic, which procedure Dr. Iago Galdston of the New York Academy of Medicine cited as The Clinical Bull in the Ecological China-Shop.

Mr. Wickenden emphasizes our disregard of the insidiousness of the poisons intended for life forms other than those warm-blooded, when we forget the similarity in all life processes when reduced to the body magnitude of the single cell. The insidiousness becomes evident in breakdown of the liver, the kidneys, and the skin—our biochemical protectors—through conversions by them of poisons for their excretion but not without the price of slow degeneration of these and other organs themselves.

The reading of this book may well have an alarming effect. It will stir some thinking about independent action for your own well being, when technical manipulation of our agricultural products en route from source to consumer results all too often in decreased, and even poisoned, food values.

It should make one intolerant of "the pathetic belief that plants, animals and humans must be dosed and 'protected' if they are to be 'healthy,' this complacent toleration of any commercial innovation so long as it does not actually kill any one (and even sometimes if it does), which is taking us further and further away from better husbandry and better nutrition. It is establishing invalidism with all its costly apparatus of treatment, as a normal condition. It should reaffirm our own conviction that "preliminary inquiries are preferable to postmortem inquests."

You may find endorsement of the book for your reading of it in the fact that three doctors of medicine contribute, respectively, the foreword, the preface and the introduction. These initial eight pages will deserve re-reading after the book has been read.

For any of us concerned with soil as nutrition via agricultural production this book is a challenge to our thinking. It is by no means a collection of facts for complacent acceptance.

OUR SOILS AFFECT NUTRITION

The twentieth century will be credited with the addition of the science of nutrition as a major contribution to the better life of our people.

Better nutrition is leading us to think less about medicine as cures and less about fighting microbes with drugs. In a more positive way, it is helping us to think more about helping the body defend itself by being well-fed and therefore healthy.

If we are to bring about good nutrition by means of good food, to build up a good defense for the body, that defense must be strong, not only against enemy invasions—as it can be against tuberculosis—but also against the degenerative diseases like heart trouble, cancer, diabetes.

The science of the soil and its fertility, by which alone high quality foods can be provided, may well be an addition during the present century to our knowledge of the better functions and better health of our bodies.

It is proposed, therefore, in this discussion to lead you to think about the health condition of our bodies as they are related to the fertility of our soils.

In considering soil fertility as it yields excessive carbohydrates but deficiencies of proteins and minerals, we need only look at the chemical composition of the human body in comparison with that of plants.

From these analytical data we see that potassium is taken into the plants in largest amounts of all the mineral elements from the soil, while calcium and phosphorus are next in that order. In the human body, these same elements are the major three, but calcium is first, phosphorus second, and potassium third.

Of amounts still higher than any of these in the human body is nitrogen. This is the key element distinguishing protein synthesized as amino acids from the elements only by plants.

Plants offer us mainly carbohydrates with only small amounts of proteins. Plant composition, considered as our food, represents possible shortages of proteins, of calcium, of phosphorus, and of probably other essential elements. We, like other animals, are constantly in danger of deficiencies of proteins and minerals, especially as we are more vegetarian.

By the very nature of the creative processes that start with the soil, carbohydrates are plentiful while there are deficiencies of minerals and proteins. Man is, therefore, always faced with shortages of minerals and proteins relative to the carbohydrates and fats. It is this nutritional need that encourages his carnivorousness and his use of animal products such as eggs and milk.

To the soil chemist, that which may be interpreted as merely allegory in scriptural language is in reality a great truth.

Viewed in chemical terms, the adult human body of 150 pounds contains only about 5.5 pounds of ash—the noncombustibles that came from the soil. This is the handful of clay into which all the processes of creation serve to blow the warm breath of sunshine, of water, of air and all else from above the soil, to build our bodies by way of the foods we eat.

Many are the failing bodies that reveal weaknesses in their structure and in their functions.

Medical science is moving from cure to prevention. Many leading physicians point to poor bone structure—commonly called rickets—and to disturbed body functions under the more understandable term of malnutrition, traceable back to the soil.

Conservationists are joining with them and going even a bit farther back to fundamentals in thinking and searching for the causes of poor body structure in deficiencies in the very handful of clay from which we are made.

SOILS, PLANTS AND NUTRITION

In considering the theme of *Lifelines of America*, one is reminded that in any rescue by a lifeline the fixed—or secure—end determines the very possibility of rescue. And in agriculture, the secure end is always the creative or the productive end; namely, the soil and its power to keep the creation of life going. While we may balance the several parts of farming against each other, the soil is the fulcrum which must hold all phases of farming high enough above subsistence if any idea of balancing one against the other is to be entertained.

We are now turning more attention to the bigger and better yields per worker as the key point under agricultural production. The bigger yields contribute to the economic line of life. But it is the higher quality as nutrition that contributes to the lifeline in its fullest meaning of life. Food for health is now getting consideration along with food for satiation.

Within the food lot, or supply, we can readily grow the carbohydrate crops, the fattening crops, and those in large surplus quantities. But the protein crops have always been considered so "hard to grow," since so much fertility must be added to the soil in order to grow them.

Crops are struggling to grow themselves rich in proteins. When those growing more proteins fail, we accept crop substitutes which grow less protein.

Some Missouri corn samples, 1955 crop, tested for protein were as low as 5.15% of crude production. That figure is just 50% of the 10.3% protein quoted as the average of Dent corn in Henry's *Feeds and Feeding*, published in 1911. Wheat has shown its decline in protein content, too. Survival of successions of different crops on the soil, given no uplift in its fertility, means that we are accepting crops with less protein.

Those going lower in protein are less able to protect themselves against fungus diseases and attacks by insects. They are also less able to provide vigorous seed by which the next generation is to be produced and to survive. Healthy plants seem to use proteins for protection.

Animals are constantly searching for more protein, too. They demonstrate their unique ability as connoisseurs of feed quality when they select the haystacks grown on fertilized soil, remain more fecund, and give different quality in wool, for example, when fed forages from more fertile soil.

Humans are not yet so capable in selecting food for quality in terms of nutritional values contributing to better health in terms of good growth, ready protection against microbial invasion of the body (disease), and prevention of sterility.

We are slowly realizing that the lifeline of America is the foodline with the fertile soils for protein production as the security point from which that lifeline reaches out.

HEALTH, AS DIFFERENT SOIL AREAS NOURISH IT

It is at the supermarket that we get most of our food. There we come face to face with the prominently posted purchase price of each item. There the amount of food we take home is limited, all too often, by the funds which our budget allows. Accordingly, we cut ourselves short on quantity. But here we are considering food only in the light of its function of relieving hunger. Unfortunately this allows for little concern about its nutritional value.

Proper nutrition provides health which allows us to serve as useful, constructive citizens in the community and the country. But nutrition to keep health at that high level must look far beyond the supermarket and price tags. It must see healthy animals for our meats, healthy plants as their feed and as our food. These require fertility in rock minerals and organic matter, providing health from the ground up in suitable geoclimatic settings. We need to appreciate the remark of a geologist who said, "We are what we are because of where we are." He was looking beyond the soil. He was including, in our nutrition, even the very rocks of the earth's crust from which the varied climates gave us varied soils in creative nutrition for life forms.

For the last couple of decades, our studies of agriculture have not been given so much to soil as to the economics of it. This looks as if agriculture were as simple as any industry, say manufacturing washing machines. But agriculture, quite differently from industry, is first the creative work of nature, by which growth of life processes is nurtured. It is not compelled or driven by man. It is only being hoped for through faith in the season, the soil, the seed and other natural factors.

The high percentage of rejections of inductees into the armed forces indicates a high degree of correlation between condition of soils and health of plants, animals and humans.

Defective teeth seem to be the most universal ailment of the human body. Animals are likewise affected. An interesting study of the health of teeth was carried out in 1949 and 1951 by Abraham E. Nizel, D.M.D., and Robert S. Harris, Ph.D., from Tufts Dental School, Boston, and Massachusetts Institute of Technology, Cambridge, respectively![1] connecting the health of teeth with the soil which grows the food. The development of caries in hamsters was the quantitative testing tool.

The foods included cornmeal and dried whole milk. In the first part of the test, these products were obtained from the state of Texas, specifically from the Panhandle, near Deaf Smith County and its town of Hereford, and also from Lubbock. Not too long ago Hereford was in the news as the Town Without a Toothache. This is the area of red and yellow soils, and of excellent cattle growth. The annual rainfall is near 25 inches. The soil shows neither leaching nor accumulation of alkalinity, as the precipitation and evaporation are approximately in balance.

In the second part of the test the corn and milk come from Orono, Maine, and St. Albans, Vermont, respectively. The soils there are said to be "podsolized," or highly mature as living soils. Annual rainfall is somewhere under 40 inches, a precipitation much above annual evaporation at that northern latitude.

Alfalfa, also a part of the food used in the experiment, was obtained from

Ithaca, New York, for both tests. The rations were made up of corn, 63%; dried, whole-milk weanling powder, 30%; alfalfa, 6%; sodium chloride, 1%; and distilled water.

Forty weanling hamsters (21 males and 19 females) composed the lot on Diet No. 1, the corn and milk from New England. Thirty-seven hamsters were started on Diet No. 2, the corn and milk from Texas. "Weighings were made weekly on each animal. After 15, 16, 17, 18, 21 and 28 weeks, representative hamsters (a total of 25 from each group) were autopsied. The teeth were examined, scored and charted according to the Keyes procedure.[2] The scoring was spot-checked by others with experience with this procedure."

They reported as follows: "The hamsters gained approximately 5 grams weekly on both diets. Caries was noted in the teeth of hamsters from both groups at the first autopsy (15th week): Male hamsters were more susceptible to caries than female hamsters; maxillary teeth were more carious than mandibular teeth; and there was a very striking preponderance of caries in the occlusal pits of the maxillary molars, particularly the second molars. Only two hamsters (both on the Texas diet) were caries-free.

Detailed data concerning percentages show that "there were approximately 40% more carious teeth, and approximately 50% more carious surfaces, in the group fed New England corn and milk. Both these differences were studied statistically and were found to be significant at the 1% level. It is evident that the area in which foods are grown may affect the prevalence of dental caries."

These data reinforce what nearly 70,000 inductees into the Navy for the Second World War revealed.[3] Their average age was 24 years. The mean data by longitudinal soil belts two states wide, coming out of the midcontinent west of the Mississippi River, were: cavities, 8.36; fillings 3.72; total cavities, 12.08. For the soil belt two states wide along the eastern seashore, the corresponding figures were 11.45, 6.10 and 17.5 respectively. Accordingly, inductees from New England, which falls into the highly-weathered soil belt of "murmuring pines and hemlocks," had 45% more cavities than the inductees from Texas, where the soils are of lesser degrees of development under moderate rainfalls, with prairies and plains. For New England alone, the data in the same order as above were: cavities, 13.5; fillings 7.8, a total of 21.30 cavities or over 75% higher than in the inductees from Texas.

These are facts which clearly tell us that the human is another biotic specimen, and our health, like that of other strata below us in the biotic pyramid, is different according as the different soil areas under the various geoclimatic settings nourish it. Apparently man is not yet "managing" his own health nutrition-wise as well as nature did that of the strata below him.

With the glib tongue of a memorized recitation, we say that good nutrition requires carbohydrates, proteins, fats, minerals and vitamins. Then we list the foods to supply these factors and assume that by assembling these foods we have compounded a ration complete for healthy growth. Courses in feeds and feeding of livestock have been taught as if the task were that simple.

We say, "Milk is a complete food; corn is a good source of carbohydrate; legume hays can supply the proteins, some of the vitamins and minerals; corn carries fats." On such knowledge we start out to fatten pigs and calves—mainly castrated males. But during many past years no attention was paid to the unknown factors in nutrition; to the differences in nutritional values of feeds and foods dependent on the fertility differences in the soil on which the vegetation was grown. Considering only the inorganic elements coming from the soil, we have not yet catalogued all the essential ones with a high degree of certainty.

Our belief in the essentiality of more soil-borne elements should be encouraged whenever any one occurs in the blood of animals and man; and, also when it occurs in the milk of any of the mammals. "Whole milk," reports John J. Miller,[4] "contains all the vitamins, minerals and amino acids officially admitted to be essential in human nutrition; but also additional vitamins and minerals whose essentiality is not generally recognized. For example, folic acid, pantothenic acid and vitamin E, among the minerals. The amino acids include the so-called nonessential as well as the essential—indicating that the body prefers to ingest them together, because of their synergistic action in the synthesis of protein.

"Note well that some of the minerals in the best quality of milk are present in very small, or so-called 'trace' amounts and this has raised a question in the minds of some nutritionists in years gone by as to the essentiality of such elements, e.g., cobalt, copper, manganese, molybdenum, zinc and even magnesium." We might well ask whether one could not apply to eggs what has just been said for blood and for whole, fresh and unheated milk.

Above, were reported experimental tests by Doctors Nizel and Harris, which in 1950 revealed that hamsters fed on corn and milk produced in Texas showed decidedly less caries in their teeth than those fed on corn and milk from New England, even when the same alfalfa and salt were supplements to both rations.

This study was followed closely by a second one to learn whether the cariogenic factor was in the corn or the milk or both.[5] The details of this procedure duplicated those of the previous tests. The study included: Group A, hamsters given New England corn and New England dried whole milk; Group B, given Texas corn and Texas milk. There were added the groups with interchanges of the sources of corn and milk. Namely: Group C, given New England corn and Texas milk; and Group D, given Texas corn and New England milk, again supplemented by a common alfalfa from New York and the ordinary salt.

Again, the hamsters were sacrificed at the end of the test period, and the dental caries scored according to the Keyes procedure. The data for groups A and B were a close duplication of those obtained in the first test reported in this section. Accordingly, this second trial showed again 13.1% decrease in carious molars and 50.9% decrease in carious surfaces for Texas-fed hamsters over the New England-fed hamsters. This agreed very well with the corresponding values, namely 41% and 47%, respectively, obtained in the first test.

However the data for groups C and D showed differences of but 6.4% for the number of carious teeth and only 2.4% differences for the carious surfaces. "There was no statistically significant difference in the caries index. This indicates that neither the Texas corn alone nor the Texas milk alone inhibited the dental caries."

"Diets A and B were analyzed for 13 nutrients, and only the riboflavin was found to be significantly different; it was approximately 0.30% higher in Diet A.

"It appears from these results that the caries-promoting property of the New England foods outweighed any caries-inhibiting property of the Texas corn and milk, and that a mechanism operated in the New England foods which aided materially in production of dental decay."

Stated more simply, the nutritional requisites for healthy bodies were not met by a fragmented ration assembled from foods grown only in part—either for carbohydrate or protein on the highly calcareous and phosphatic soils of the Texas area near the town of Hereford. Such rations did not prevent degeneration of the enamel structure of the teeth. This reports nothing about the probable disturbance of other metabolic aspects. Is the nutritional failure for natural self-protection apt to be due to a single positive factor or mechanism of degeneration? Isn't it more apt to be caused by an absence, or even by an irregularity, in a chain of many contributors to self-protection?

The training and education of professionals ministering to our ailing health seem to encourage search for only positive, damaging factors. Seemingly, dental caries should invite thinking about missing factors, or in the negative, the failing maintenance of the high level of our biochemical complexity guaranteeing good health. Of course, thinking of a single factor means single cause and single cure. This is inviting for prescription of a single drug or single element, like fluorine.

But in the naturally healthy setting for cattle in Texas and in the good human teeth discovered there, that "single-factor" thinking is apt to be erroneous when it attributes good teeth to the fluorine from the surface soil going into the drinking water en route to the sea. That ionic fluorine comes from the natural weathering of the magmatic mineral, apatite. While activating the soluble fluorine, it is simultaneously releasing to the surface soil calcium and phosphorus in more available forms for the better nutrition of plants and animals. For this reason there is better bone-building in that area, according to a report by Lewis B. Barnett, M.D., Dallas, Texas.[6] Doesn't it suggest erroneous reasoning to ascribe better teeth and bones to one anionic, highly corrosive element in the drinking water than to the two cationic ones, calcium and phosphorus, the most abundant nutritive elements in the ash of the warm-blooded body? The prize Hereford cattle, drinking surface water, gave the town its name long before the newspaper publicity characterized it "The Town Without a Toothache."

Single organic factors in nutrition, like either Texas milk alone, or Texas corn alone, did not prevent or eliminate dental caries, but moved decidedly in that direction when working together. This said nothing pro or con about including alfalfa from New York, as was done in the scientific tests by Doctors Nizel and Harris. In dealing with the biochemical complexity of healthy teeth, does not a single factor, like fluorine in our drinking water, seem a highly illogical premise on which to prescribe and enforce mass medication in an area where health is failing?

SOIL CALCIUM AND THE QUALITY OF LEAFY GREENS

Green leafy vegetables are recognized as important foods in the human diet. As the providers of minerals and vitamins they are among the "protective" foods recommended by nutritionists. It has long been realized that there are significant and important nutritional differences between certain greens of the mustard family (kale, mustard greens, and turnip tops) and those of the goosefoot family (spinach, Swiss chard, beet greens, and New Zealand spinach).

The nutritional superiority of the vegetable greens of the mustard group must be ascribed to their higher contributions of the inorganic element, calcium, and of the vitamin C, ascorbic acid. The calcium serves in body building and the vitamin C in stimulating body processes, especially those providing energy.

Naturally, attention has been directed to the importance of soil fertility as a determiner of the "nutritional quality" in whatever we grow, whether that be crops in the garden or in the field. We need to focus our attention continually on such quality when even declining soil fertility has given us "surplus" crops in the category of those that fill, fool and fatten, but not of those that truly feed. And it is important to emphasize the calcium in "greens" when so much is said about eating them to get "minerals" or, rather, the inorganic elements for building the bones of the body.

Some experiments showing how much difference the fertility of the soil makes in the concentration of the inorganic element, calcium, were made possible by using the carefully prepared colloidal clay to supply the plant with differing amounts of calcium per plant, namely 40, 20, 10, and 5. This gave an extensive range in ratios of these two nutrients to each other. The seven kinds of vegetable greens cited above were grown, and careful chemical analyses made of them for their concentrations of calcium in the dry matter.

The concentrations of calcium, as percentage of the dry weight, are most easily shown by the curves in the accompanying chart. The superiority of the three greens, in terms of the concentration of the calcium in the crop, is well shown when the curves for kale, turnip greens, and mustard greens are so much higher on the scale at the left of the chart than are those for spinach, Swiss chard, beet greens and New Zealand spinach. The figures for the latter four did not go over 1%, yet for the former three did not fall as low as 1%, as providers of calcium in our diet. There can be a decided difference in the nutritional value of the kinds of greens one chooses.

In addition to the amounts in total by which the kinds of greens are different, it is more significant to note the betterment in quality, as providers of calcium, according as more calcium was applied to the soil. The mustard family, which was superior in its delivery of more calcium in total, also increased its delivery of nutritional value in this respect more than did the goosefoot family when more calcium was added to the soil. This is evident from the greater angle in rise in the upper three lines than in the lower four lines in the chart (see chart on page 245).

The curve for kale shows the most pronounced rise for the concentration of its calcium with more calcium in the soil. Some figures for the calcium of the better kale crop showed how wide these differences were.

When 40 units of nitrogen were applied in the soil and combined with 5, 10, 20, and 40 units of calcium applied, the yields of crops were respectively 558, 468, 544 and 523 grams to make their appearances too much alike for anyone to distinguish between the similar plants had they been put out for your discrimination. But yet the calcium concentrations were respectively, 1.98, 2.25, 2.46, and 3.10%, as more calcium was put into the soil—an increase of over one-half by the highest over the lowest.

This is a decidedly higher delivery of calcium per mouthful of greens according as the soil was treated with more calcium to bring about this greater nutritional contribution by the crop. The choice of greens can improve the nutritional values put into our diet, but those values are also raised by the uplift of the fertility of the soil with respect to the inorganic nutritional element in question.

These studies, like many others, point out what we can do for the improvement of our nutrition. We should not be satisfied to think only the names of the variety or species of plants we grew, but should also concern ourselves about the fertility of the soil under the plants . . . when this starting point of all plant growth determines what nutritional qualities will be created by that natural performance.

PROTEIN SERVICE IN NUTRITION

Nutrition has done much to interpret the fuel and energy values of feeds and foods. Plant physiology is interpreting plant nutrition for carbohydrate delivery as this registers changes in plant bulk, by more or less tons and bushels. Unfortunately, in the nutrition of both plants and animals, we know all too little about the roles played by the proteins; the foods that rebuild the body, that carry life and that guarantee reproduction. They, in certain liquid forms as serum and in compounds of cellular dimension as corpuscles, constitute the bloodstream.

It is those proteins that combat invading microbes, that build antibodies and give protection against so-called "disease," "allergies," etc., in biochemical ways and means yet unknown.

The myriads of different kinds of proteins coming about by no more than just rearrangement and varied combinations of the constituent amino acids of that molecule, are still in the realm of mystery.

Undergirding the bloodstream, through the synthesis by the body of the special proteins for it, is now being considered a significant role of good nutrition when we speak of "protective" foods. Through the wider acceptance of that principle, a big step will be made toward the absence of disease and the presence of its counterpart, namely, good health.

Some recent tests of alfalfa for its content of the different essential amino acids, as related to soil treatments with trace elements, pointed to deficiencies in these components of protein according to the trace elements as soil deficiencies.

These demonstrations suggested the working hypothesis that possibly the declining and exhausted supplies of soil fertility are responsible for less synthesis by crops of proteins in total (so crudely measured in terms of total nitrogen multiplied by 6.25) and for less, or absence of some specific amino acids. With total protein going lower in our wheat; with a drop from 9.5 to 8.5% of it in corn during 10 years; and with the deficient specific amino acids in corn possibly low because trace elements are deficient, should we not turn to considering some diseases as possible deficiencies coming by way of the soil?

Isn't it good nutrition that is used as the "cure" for human tuberculosis? In that "disease" the effort is not given to the extermination of the microbes from the lungs and other body parts by means of antiseptics and other sterilizing agents. Instead it is nutrition by milk, eggs, meat, and all else for a high protein diet.

Under such treatment, the germs apparently recognize their premature anticipation of a task of disposition and literally move out. Shall we emphasize the "cure" in this case or shall we raise the question of whether deficient nutrition and defective physiology were in advance of, and an invitation to, the entrance by the microbes? Were the "germs" the cause then, or merely an accompanying phenomenon of what is a deficiency but which we call tuberculosis? Might this not be the cause for some of our cattle diseases, accompanied by microbes, but yet so baffling that slaughter is still the "cure." In cases of undiagnosable animal ailments, the able veterinarian often recommends feeding good alfalfa hay grown on the more fertile midwestern soils, or he prescribes some extra amounts of other protein supplements, as accompaniments to his medication. When the animal recovers, a similar confusion as to correct explanation of causes for the animal recovery is involved.

SELF-PROTECTION BY PLANTS
LINKED TO NUTRITIONAL VALUES

In previous sections, the quality of food plants in terms of nutritional values was discussed. Crops classified as vegetable "greens" were considered, with spinach being singled out for detailed listings of varied chemical compositions in relation to some of the factors controlling different qualities.

The modifying factors are numerous enough to warrant their discussion in order to comprehend and appreciate the refined combinations in "natural balance" by which nature has developed healthy plants during their evolutionary ages for man to domesticate. More appreciation is demanded, now that our management of crops has made them sickly, allowed them to be taken by pests and diseases, and in view of our controversial views of qualities of a vegetable, like spinach, which has both friends and foes concerning its fitness as food.

We need to study nature more, and books less, as Agassiz, one of our great scientists, suggested about a century ago. Nature was in the worldwide business of growing plants without treatments to pamper the soil or protect plant tops against pests many years before man's accumulated and recorded experience can cite claims for his successes in that productive activity. We are gradually returning to nature with confessions, as prodigals, that we are not able to duplicate her successes in climax crops. More of us are attempting to duplicate nature's methods, rather than continue in our efforts to fight pests and diseases with more and more powerful poisons.

Mentally stampeded by profit motives, we consider these seasonal damages to be caused by invading enemies. It would be more logical to view them as nature's elimination of plants in poor health, unable to survive because of our omission and disregard of factors which would keep them "well-fed and healthy" and able to exercise the same degree of self-protection that they possessed before man's arrival.

When vegetable crops must be propped up by sprays of highly poisonous, polyphenolic and aliphatic compounds just to carry them near enough to maturity to be harvested, shall we not consider them of low nutritional quality simply because they were not healthy? Since plants take in many complex organic compounds via leaves and roots, shall we not consider the danger in the crop's absorption and retention of sprays which are highly lethal to humans as well as to insects?

Nature has been unique in balancing plant diets. She used no soluble salts. She grew crops on soils where reserve rock minerals were decomposing through microbial help from decayed organic matter. This growth accumulated to bring about inorgano-organic combinations of elements and compounds for balanced plant nutrition by selection and synthesis from the same kind of preceding crops. Now that we are testing soil to detect the first limiting element, which we then apply, then the second limiting one, and so on, we are bringing about an imbalance of salts, rather than a balance as achieved by the natural plan of plant nutrition through balance of elements and all other factors in the combination of weathering minerals and decaying organic matter from which solubles are leached out to sea.

Spinach, like almost any other crop, is highly responsive to nitrogen applied only to the soil—even if such treatment gives imbalances in the nutritional values of the crop as food. When researchers at the Missouri Experiment Station varied the nitrogen applied to the soil to study the effect of that factor on the concentration of the oxalate and vitamin C, the data revealed some interesting facts. (See figure 2 on page 112).

Among these was the rapid increase of the crop yield with increments of applied nitrogen, starting at 5 milligram equivalents (ME) per plant and doubling to give the series 5, 10, 20 and 40 ME as the soil treatments. but more significant was the decrease in oxalate concentration (from a high of near 8.15% to 6.25%) with the increase in applied nitrogen from 5 to 40 ME per plant. When nitrogen was applied in these amounts, the first increment resulted in an increase of vitamin C concentration amounting to 0.5 milligrams per 100 grams. Then the vitamin C dropped from a high of 29 milligrams per 100 grams to a low of near 21.

This suggests that vitamin C is a catalyst in the plant's life and is at high concentrations for stimulation of nutritional processes, but at lower concentration when more nearly balanced nutrition allows these processes to be more normal under less stimulation. It follows that it would be an error to try to grow a crop for high vitamin C if the content of this factor is high only when the plant process is abnormal. But, as seed and vitamin storage for starting life anew, fruit high in vitamin C would, we believe, be most healthy—quite the opposite of that for vegetative matter.

The most interesting fact about quality revealed by the experiment was a result of accident: an attack on the spinach crop by the leaf-eating thrips insects (*Heliothrips haemorrhoidalis*). The leaf damage they created decreased from 40% at treatments of 5 and 10 ME of nitrogen to near 10% at 40 ME. This suggests that with nitrogen as a factor in protein synthesis, there is not only an increase in nitrogen's support of growth of living tissue, but also a synthesis of compounds for self-protection. Also, there is less metabolic waste if the oxalate reduction is correctly interpreted when put into that category.

All this indicated that proper protein synthesis by plants given a balanced nutrition through balanced soil fertility results in "healthier" plants, more fit to survive attack by pests, and providing more healthful food for human consumption.

EAR, NOSE AND THROAT DYSFUNCTIONS
DUE TO DEFICIENCIES AND IMBALANCES

When the unusual title of this book tells us that dysfunctions of such vital organs as ear, nose and throat are due to deficiencies and imbalances in cellular nutrition and metabolism, there is a decided lift in the hopes for anyone's health through his own care of it via nutrition. Dr. Sam E. Roberts, who has been demonstrating these facts in his own practice as a physician, is therefore speaking forcefully when he says, "Emphasis on disease, often to the unfortunate exclusion of the consideration of health, has dominated medicine." "Physicians," he says "are so busy treating acute conditions, they fail in their duty to the patient whose only complaint is 'I don't feel well. I need a check-up.' It is so much easier and less time consuming to tell those patients, 'There is absolutely nothing physically wrong with you,' leaving the distinct connotation that the condition is nervous or mental."

As the result of convictions resulting from nearly 45 years of experience as a teacher in a large medical school and of close to 50 years in private practice, Dr. Roberts courageously prefaces his volume by saying, "I am certain that deficiencies, and especially imbalances, are the principal causes of the diseases, dysfunctions and syndromes herein reported. They are the contributing etiologic factors to many other diseases and dysfunctions not included in the text. They are the precursors of many chronic degenerative diseases. These deficiencies and imbalances are nutritional, electrolytic, acid-base and insulin-sugar."

"Through correction of these deficiencies and imbalances, when indicated, the symptoms and physical findings are relieved or greatly improved. This I will call throughout the volume, the therapeutic test.

"Available laboratory tests have, in the main, impeded, not helped, these investigations."

On the basis of careful, and even presumptive, diagnosis in attempting to locate the deficiencies and imbalances in the patient's nutrition and metabolism, Dr. Roberts has been an outstanding scholar. He has had marked success in establishing health via applied knowledge of biochemistry and therapeutic nutrition.

His studies reported in this book include the major inorganic essential elements, (coming from the soil through plants) hormones, vitamins, and other items of recently recognized deficiencies and imbalances, to say nothing of excesses in many common habits.

With careful discussion, at the outset, of the broad principle of dysfunctions as these troubles are based on deficiencies and imbalances, the volume then devotes its major part to classifying headaches, troubles in hearing (Miniere's Disease), allergies, nasal or sinus dysfunctions, throat troubles (Hilger's Syndrome, Sclerosis) and many others in this category; and to giving details of the basic therapy for restoration to proper function in these many cases. Since most of this recovery depends upon the patient's effort and cooperation, including decided changes in perhaps well-established habits of eating or drinking and the use of drugs, etc. Dr. Roberts' book becomes the physician's means of teaching the patients their own responsibility for their own health. In this he has been a great and beloved teacher, giving separates of printed instructions through the years, while this more complete and printed report on his methods comes forth under its unique title at just about the date of his retirement from this phase of his life's work.

As a scientist, Dr. Roberts pushed his curiosity, observations, hypothesis and conclusions well into the medical frontier of human health. He followed through on his early conviction that "No patient should be branded with a diagnosis of a neurosis or psychosis until the physician has made a systematic and vigorous attempt to correct all known and even suspected deficiencies and imbalances." He confesses, that, "The opinions and annotations herein recorded are founded only on clinical observations. I do not claim originality for anything and willingly admit the only 'documentation' is my memory of events and what I have retained from a constant perusal of medical literature for nearly half a century."

"I have used numerous working hypotheses to explain certain facts in the true sense as I have observed them, and as a guide to investigate others. There is little or no experimental evidence against these various hypotheses. In truth, most of what is reported seems to fit."

"A hypothesis today may be a fact tomorrow."

In the fraternity of his time, Dr. Roberts has been an early scientist in this subject of his choice, and an original pioneer of practices that emphasize the wholeness of the body with its refined integrations of many delicate processes giving us good health. He has approached irregularities of that condition of ourselves as if they were a disfunction, and in the belief that they are apt to be sins of both omission and commission of our own. He has attempted to help find the physiological evidence through which the patients' efforts could restore those disturbed processes and conditions. Dr. Roberts' book is proof that he is a great humanitarian as well as a scientist and a teacher, whose recent retirement comes all too soon in the light of the services he has been rendering. His life has given new light on better health and longer life as a matter of proper nutrition.

8

PRAGMATIC ALBRECHT

THINK OF THE SOIL AS A FACTORY—NOT AS A COMMODITY

Let us imagine a small water power electric plant with a regular and continuous production of 1,000 watts of electric power. The entire electrical output is sent to an adjacent manufacturing plant. A special product is manufactured in this factory. The machinery in the plant cannot be regulated and it runs at full speed continuously. If the owner supplies the right quantity of raw materials, almost like magic, 20 times as much material comes out in the finished product. Let us say 50 pounds of raw material starts through the factory and we produce 1,000 pounds of finished products. The proportion must be correct. Especially if we send too little raw material through, the faulty finished product has little or no value. Visitors at the manufacturing plant often see the machinery running at full speed, but no raw materials being sent through. "What a waste of power," they say. "Think what could be produced if this plant was operated efficiently and the owner would only provide the small amount of necessary materials."

Let us compare this imaginary factory to the soil. The soil is a factory as well as a stock of raw materials. The soil must convert the more resistant plant nutrient elements into such forms as are taken by the living plants and at such rates as will produce significant plant bulk for an economical crop. Sunshine is the power which operates the soil factory. The soil's contribution to the vegetation is normally about 5% and occasionally as high as 10%. The atmosphere normally contributes over 90% of the total dry weight. This inflow of air and water gives us that mystical performance we call plant growth. In a typical example, the soil contributes but 50 pounds per 1,000 pounds of dry vegetation matter. Like the inefficient manufacturing plant, unless these raw materials of the soil are delivered promptly the remaining 950 pounds will not be captured from the air and water. Therefore, the sunshine power will remain unused. The unproductive crop field, the field depleted in minerals, is like the manufacturing plant with its machinery running at full speed and no products being produced.

Let us consider the stock of raw materials in the soil as being like a store with its shelves containing the stock of goods. Each sale of farm products should be considered as a removal of fertility elements from these soil "shelves." A practical example illustrating this analysis is as follows:

A 200 acre farm of virgin Marshall silt loam was divided into five 40 acre units bearing corn, oats, wheat, clover, and grass. A five year rotation gave the following yields per acre annually: 50 bushels of corn, 35 bushels of oats, 20 bushels of wheat, 2 tons of hay, and 1.5 tons of grass hay equivalent. These yields would have the following monetary values:

	Five Year Total Pounds	Annual Pounds	Value
Nitrogen 191	38.2	$3.82
Phosphorus 43.5	8.7	1.30
Potassium 187.6	37.5	1.87
Calcium 106.3	21.2	0.02
Magnesium 39.7	7.9	1.12
Sulfur 9.5	1.9	0.02
Iron 6.3	1.2	
Sodium 5.2	1.0	
Silicon 106.0	21.2	
Annual total value per acre $8.15		

No value assigned to iron or silicon in this table. They are important essentials in food.

Have you ever imagined that on a 200 acre farm the fertility equivalent of $1,630.00 is taken out of the soil annually? The fertility removal as indicated in the above chart in terms of respective nutrients would cost that amount at the commercial fertilizer prices. This $8.15 an acre represents an annual production of 3,824 pounds of organic output at the cost of 200 pounds of the soil commodity. One pound of the soil is expended for 19 pounds of produce grown. In terms of fertilizer prices for the soil expended, the organic matter produce is costing more than 20 cents per hundred weight. A rental of $4 per acre for this land would replace less than half the fertility coming out of the soil.

If this farm was used for the production of animals for fattening or working purposes, the loss would not be so great. Nearly 85% of the fertility ingested in such farm animals is excreted. Then, if $1,630 worth of fertility capital all goes through such animals, more plant nutrients are required for growth and therefore less excretion. In this case, the excretion would be about 65%, with a fertility value returning to the soil of approximately $1,060. In the first case, an inventory of the soil fertility loss would amount to $245 and the latter case $570 annually. The actual loss will be greater than this amount since, because of carelessness in handling manure, a minimum loss is estimated at 25% of the total value of the manure.

Should any grain or forage be sold rather than fed, the soil fertility suffers additional losses. The loss of soil nutrients from an orchard crop is even greater than one would at first believe. An apple orchard with thirty large trees per acre represented a value of $2.74 per acre in terms of only three of the major plant nutrients removed from the soil. For a peach grower with smaller trees and 100 per acre, the balance sheet would show a greater demand upon soil minerals and a total annual rotating capital of $1,306. In this case, the removal is greater than the apple grower's or the general farmer's, and again 45% of the rotating capital goes off the soil. There may be methods of producing farm products of such nature and selling specific parts so that little or no fertility would be lost. These parts would necessarily be such chemical composition containing little more than carbon, oxygen, and hydrogen originating in the air and water. Should only butter be sold, for example, it would sell no soil fertility, save that in the casein and in solution in the water in the butter. Therefore, we must carefully analyze the farm products which we sell. If they are mainly carbohydrates, they take less fertility from the farm. If they are proteinaceous substances, the drain on the soil fertility is much greater.

Since the fertility from the soil "shelves" is being depleted by each crop, it is important to know how long this may continue. Should we take the simple calculation that about 50% of the rotating fertility is being removed annually, it is still necessary to study this in relation to the entire stock in the soil. Using Marshall silt loam with 3,800 pounds nitrogen per acre in virgin surface soil, the rotation mentioned in the above would use only one-half of 1% of the original supply. Using phosphorus at 1,200 pounds in the virgin soil, the loss was 2.9 pounds annually or more than one-fifth of 1% of the total. The nitrogen supply would be completely exhausted in 200 years. Using the soil as a commodity, it would be "sold out" in 200 years.

We dare not focus attention on any particular soil element. Plant tissue needs eleven mineral salts (containing sixteen elements) a supply of carbohydrate, three vitamins and one amino acid. Omission of some like magnesium, calcium or sugar causes immediate stoppage of growth. Lack of others results in abnormal growth. Probably sixteen elements are demanded for plant construction and at least sixteen are required for animal body building. Four of these, namely, carbon, oxygen, hydrogen and nitrogen constitute the bulk of all plant and animal forms. These are not of soil origin but are supplied by air and water. The plant draws ten of the above from the soil. They may exist in extremely small quantities in the plant or animal. However, life is impossible with any one absent. It is thus in terms of absolute supplies of body building nutrients that the soil controls and directs the pattern of life. The soil limits life to those forms and numbers that can be maintained by the nutrient supply delivered by the soil.

In our study of the soil, we need to find out what it is that makes the soil fertile. The farmer may sell a variety of products but they represent one thing— the fertility of the topsoil of the field. From the above discussion, it is apparent that the average farm is decreasing in fertility. However, we should point out that this need not be, since we find some soils in China, Egypt, and other parts of the world through proper cultivation have long maintained their fertility at a high level. An important factor in maintaining this high fertility is the return of organic matter to the soil. [Our failure to do this is discussed in another section of this report.]

As the soil loses its fertility, the type of plants which will grow also changes. Soils rich in minerals may successfully produce clover and other plants having a high mineral requirement. The vegetation rich in minerals is also high in protein. The ash from this vegetation has a greater mineral content and a smaller proportion of silica. As the soils become less fertile, the vegetation decreases in mineral content and increases the proportion of silica as shown in an analysis of the ash. These plants become more woody in their composition. Millions of acres, which formerly would produce good clover crops, are now suitable only for woody plants. Our forests are located on soils which are low in minerals. Woods or timber are the last stand by vegetation against the flow of soil fertility to the sea.

The forage for grazing horses serves two body functions. One of these consists of supplying the materials by means of which the body is constructed. The other is that of providing energy to run the body machinery and move the animal and its load about. Protein is necessary in the growing animal to build muscle. It demands lime and phosphate to build the bones. Only a proteinaceous substance already fabricated by the plant can meet the protein requirement for horses or any other growing animal. They cannot use the simpler elements for making their protein as in the case of plants. The lime and phosphate also come by way of the plants from the soil. The minerals and proteins are required in large amounts by young horses, and by mares during gestation and by stallions for most successful service, because these are the "grow" foods and come from the soil.

Energy foods are supplied in the form of carbohydrates and fats. It is the carbohydrates that make up the larger part of the bulk of plants. Starchy grains and much of the plant's fibrous structure are the horse's energy or power sources that do not come directly from the soil. They are the "go" foods that plants seemingly take from the weather or those materials present without the soil. Plants won't render this service of snatching the sun's energy except through the help of 5 or 10% of their own "grow" foods that are taken by their roots from within the soil. They represent the plant's mineral part or ash. Thus horses are power plants releasing, for our service, the sun's energy stored by the plants. Thus the soil is the foundation of our farm power plant when we use horses.

As soils differ in clay content, in humus content, or in mineral reserve of the silt and sand, necessarily the crop growth will be different. As the reserves are exhausted, the kind and quality of the vegetation will differ.

Alfalfa hay is at the top of the list for its concentration of minerals and heavy demands upon the soil. Alfalfa, clover, and other legumes are universally accepted as effective feed for colts and other growing animals. Clover grows only on soils rich in lime and phosphate and is thus important growth food. The bluegrass area around Lexington, Kentucky, usually provides the winners at Churchill Downs. Success in growing these fine horses cannot be divorced from lime, phosphate, and generally good fertility in the soil. Quality of forage is more than a trademark stamped on a package. It must be grown into the goods by way of the soil.

Soils may be depleted either by man or by nature. The depleted soils result in more fuel value in the crop but less help in animal growth. Soils on cultivated lands are partially depleted through the process of weathering. The areas where rainfall is heavier, and the temperatures higher are more likely to be depleted. Calcium and other elements are dissolved in the soil moisture and lost through the process of leaching. Overcropping is another important factor in depleting soils. Probably the most important factor, however, is erosion; and it has been estimated that this loss is 20 times as great as that loss through the utilization of plants.

Many soils, having lost their fertility, will respond to proper treatment and again become productive. Such treatment includes manures, limestone, and fertilizers. Pasture research is going forward to give us better pasture. A search is being made for substitute grazing crops. We have our choice of putting additional minerals in fertilizer and organic matter back in the soil or selecting substitute crops because of their failure to do so. A study of the percentage elements in soils, plants, and the human body reveals the concentration which must take place. Calcium in animal ash is 40 times as concentrated as the mobile calcium in the soil, and phosphorus similarly more than 100 times as concentrated. Animals are in trouble when they are compelled to eat herbage getting little of these essentials from the soil. Animals know their forages so well that even a blind horse, according to Dr. Dodd of Ohio State University will graze to the line of the soil treatments represented by only a few hundred pounds of fertilizer.

If we were keeping a soil bank book we would record a loss of raw materials lent by the topsoil to the plants. In many areas the "soil bank" has closed. The bank has lost its chemical deposit. The child should understand, therefore, that as a crop is taken from the field each year the fertility of that soil is likely to be reduced. A farmer may produce certain types of plant life which will rebuild the soil. A recent report indicates that an experiment on corn-belt acres which had been cropped down to a production of under 40 bushels were brought up to a 70 bushel level by a single crop of sweet clover. Clover restores nitrogen to the soil. Clover also improves the structure and thus more soil moisture may become available to plants.

When the soil becomes depleted in certain minerals we should think of it as a sick soil. Depleted soils will not produce healthy plants. Plants from depleted soils will not nourish healthy animals. Sick animals will yield little or no income. A more important problem is the influence of these sick soils on your health. [How poor soils may result in poor health is discussed in another section of this report.]

SOME AIMS OF SOIL RESEARCH

Much as the definition of research depends on the person doing the defining, so the aims of research may depend on the one who is aiming.

For our purposes here, let us exclude from true research the repetition of previous researches for greater refinement, and of facts already established and reported.

Modern techniques of measurement permit refinements today which were not possible a short time ago.

For example, in animal feeding trials in low winter temperatures and summer highs, it was established long ago that the animal accommodates itself, by means of extra feed, to temperatures below 70°F. But the animal cannot accommodate itself so readily to temperatures above 70°F. Now, with modern refrigerator aids and heating apparatus, it is possible to study the animal's physiology by increments as small as a degree or two—and thus refine the basic facts of former research.

This repetition under more refinement scarcely is research unless it adds new facts and new principles, or it shifts the temperature figure previously established.

Research in its evolution of new knowledge usually includes four steps or phases—(1) observations; (2) theories prompted by those observations; (3) experimental tests of the theories; and (4) conclusions drawn from the results of the tests.

A few of the aims of some soil research—rather than what soil research in total is aiming to do—may be briefly outlined here.

For many years our soil research has aimed to understand how soil can hold plant nutrients in sufficient supply for an entire season when the same nutrient supply, if put into a solution as plant growth medium, would demand—for the necessary dilution—more water than the soil could retain within root reach. If put into solution in the water within root reach, the resulting high salt concentration would kill the seedlings.

As another aim in soil research, the dynamics of ionic adsorption and exchange suggested the theory of a contact area of the colloidal root and colloidal clay as the center of the chemo-dynamics, by which nutrients are held in quantity and exchanged to the advancing root, which is offering for them the active hydrogen originating in root respiration.

Fortunately, for these aims, some funds which had accumulated from the sales of legume inoculation, and an able young Swiss chemist reluctant to leave the United States after the expiration of his scholarship visa, were the combination to facilitate the testing.

All this told us that legumes—representing a proteinaceous root with higher adsorbing energies than the non-legume roots with less of protein—will exhaust soil-fertility to a much greater degree in a single crop than the non-legumes.

Naturally, then, the non-legumes can be grown on the same soil for many more crop successions than the legumes. Whether the plant is producing mainly carbohydrates, or is converting much of these into proteins, is significant in many aspects, connecting themselves with the suite of ions going from the soil into the roots and thereby into the plant.

As the plant produces more carbohydrate and more bulk, the efficiency of nitrogen as nutrient would be considered to go higher. According to the theory of Willcox, the maximum amount of nitrogen to be taken from an acre of soil should give over 6 tons of dry matter per acre in soybean forage; over 13 tons per acre in corn; and almost 56 tons per acre in sugar cane—all of these, 318 pounds of nitrogen.

The mineralogy of the soil is now a returning and revived research aim. Professor E.W. Hilgard, author of an early book on soils under date of 1914, recognized at that early date the mineral reserves in the silt more than he did the fertility adsorbed on the clay fraction, or the sustaining fertility of soils under a long period of regular productivity.

Aided now by progress in the science of mineralogy, the colloidal clay acid comes into crop production via its action in processing the insoluble mineral reserves for sustaining fertility as well as in the soluble compounds of the starter fertilizers it adsorbs and exchanges to the plant roots.

Just what soil research is aiming to do depends on the individual who is studying. It will depend, not on a majority rule of any scientific body assembled, or on agronomists grouped according to geographic, agricultural, or economic categories. Rather the aim of research will be higher as the vision of the researcher can be extended by more basic facts at his command for extrapolation of his thinking far enough into the unknown to bring maximum benefit to agriculture and all that is dependent on it.

AGRICULTURAL EDUCATION

The lack of agricultural policy, the prevailing government discords concerning any policy, and the need for an educated citizenry that appreciates agriculture and the soil as our basic resource supplying food, was a complicated problem a century ago. That we have no better solutions for the problem today than the Congress had for it in 1849 is a lamentable fact when we boast of the high literacy of our population.

As basis for this lament, even a century ago, there was the report in 1849 by Daniel Lee, M.D.[1] that "As a nation of farmers, is it not time that we inquire by what means, and on what terms, the fruitfulness of the earth, and the health and vigor of its invaluable products, may be forever maintained, if not forever improved?

"A governmental policy which results in impoverishing the natural fertility of land, no matter by what particular name it is called must have an end. It is only a question of time when this truly spendthrift course, this abuse of the goodness of Providence, shall meet its inevitable punishment.

"To illustrate such an important fact as well as a principle, let us suppose a farmer produces crops worth $1,000 and they cost him, including all expenses for labor, wear of implements, interest on capital, etc., $850. But it often happens, if he should undertake to replace in his cultivated fields as much of the potash, soda, magnesia, phosphorus, soluble silica, and other elements of crops, as both tillage and cropping had removed, it would cost him $175 or $200 to effect that purpose. It is only by consuming the natural fertility of the land that he has realized any profit.

"In a national point of view, all labor that impoverishes the soil is worse than thrown away. No fact in the science of political economy is more important than this . . . Thus to impoverish land is to wither the muscle of both man and beast employed in its tillage.

"If all the sheep in the United States gave us as good returns in wool for the feed consumed, as the best 100,000 do, it would add at least 60 million pounds to the annual clip of this important staple.

"In one of his letters to Sir John Sinclair, General Washington said, in substance, that, at the time he entered the public service in the War of the Revolution, his flock (about 1,000) clipped five pounds of wool per fleece. Seven years after, when he returned to his estate, his flock has so degenerated that it gave an average of one to two and one-half pounds per head which was the common yield of Virginia sheep then as it is now.

"Neither the earnest recommendations of the illustrious farmer of Mount Vernon, nor the prayers of two generations of agriculturalists, nor the painful fact that nearly all tilled lands are becoming less and less productive, could induce any legislature to foster the study of agriculture as a science. Happily, this term, when used in connection with rural affairs, is no longer the subject of ridicule. Some pains have been taken in this report, to prove that $1,000 million, judicially expended, will hardly restore the 100 million acres of partially exhausted lands of the Union to that richness of mould and the strength of fertility for permanent cropping, which they possessed in their primitive state.

"The continued fruitfulness of the earth is an interest far greater and more enduring than any form of government.

"If the 22 million people now in the United States may rightfully consume the natural fertility of one-third of the arable lands of the country, the 44 million who will be here 25 years hence may properly extinguish the productiveness of the remaining two-thirds of all American territory.

"There has been enough of the elements of bread and meat, wool and cotton, drawn from the surface of the earth, sent to London and buried in the ground or washed into the Thames, to feed and clothe the entire population of the world for a century, under a wise system of agriculture and horticulture. Down to this day, great cities have ever been the worst desolators of the earth. It is for this that they have been so frequently buried many feed beneath the rubbish of their idols of brick, stone, and mortar, to be exhumed in after ages . . . Their inhabitants violated the laws of nature which govern the health of man and secure the enduring productiveness of the soil. How few comprehend the fact that it is only the elements of bread and meat evolved during the decomposition of some vegetable or animal substance that poison the air taken into human lungs, and the water that enters the human system in daily food and drink! These generate pestilence and bring millions prematurely to their graves.

"Why should the precious atoms of potash which organized the starch in all flour, meal and potatoes consumed in the cities of the United States in the year 1850 be lost forever to the world? Can a man create a new atom of potash, or of phosphorus, when the supply fails in the soil, as fail it must under our present system of farm economy? Many a broad desert in eastern Asia once gladdened the husbandman with golden harvests. While America is the only country on the globe where every human being has enough to eat, and millions are coming here for bread, how long shall we continue to impoverish 99 acres in 100 of all that we cultivate?

"Both pestilence and famine are the offspring of ignorance. Rural science is not a mere plaything for the amusement of grown-up children. It is a new revelation of the wisdom and goodness of Providence—a humanizing power, which is destined to elevate man an immeasurable distance above his present condition. To achieve this result, the light of science must not be confined to colleges; it must enter and illuminate the dwelling of every farmer and mechanic. The knowledge of the few, no matter how profound or how brilliant, can never compensate for the loss incurred by neglecting to develop the intellect of the many. No government should be wanting in sympathy with the people, whether the object be the prevention of disease, the improvement of land, or the education of the masses. One percent of the money now annually lost by reason of popular ignorance will suffice to remove that ignorance.

"If 'knowledge is power,' ignorance is weakness; and the removal of this weakness is one of the highest duties of every republican government. Either the assessors or collectors of state and county taxes should be provided with blanks to collect useful information, as well as money, from the people.

"There appears to be no government that realizes its duty 'to promote the general welfare' by widely diffusing among its citizens a knowledge of the true principles of tillage, and by impressing upon them the obligation which every cultivator of the soil owes to posterity, not to leave the earth in a less fruitful condition than he found it."

A century ago this doctor left that suggestion which would give us a set of officials who are better informed on the problems of the callings and struggles of those from whom they are taking the moneys for government. It would generate their more sympathetic understanding of the difference between earning a living in cooperation with Mother Nature and collecting it from human nature.

BASIC FACTS OF SOIL SCIENCE

Let us outline, rather than discuss in detail, some of the basic facts connecting the nutrition of all forms of life closely with the soil; more particularly via the proteins which are becoming more costly as the fertility of our soils declines.

Our national fertility pattern, resulting from our climatic pattern, gives corresponding patterns to crops, livestock, nutrition, health and other aspects of all life whereby it puts the soil in control.

The varied concentration of farms in the different regions of the United States outlines the varying chemo-dynamics of the soil for nutrition in a pattern duplicated, for example, by that of the radio efficiency. There are more farms in the same areas where there is the better radio service. When it is the fertility salts of the soil that are both the soil's conduction of electricity and the soil's production of crops, shall we not see the variable chemo-dynamics of the elements of soil fertility responsible for both the better crops, and the better radio transmission, to say little about the better physiological settings for higher forms of life?

The national patterns of these activities in farming and radio efficiency are determined by the pattern of the climatic forces (rainfall and temperature) that have been decomposing the soil minerals and keeping the stream of nourishment flowing to give us our feeds and foods of higher protein values for better maintenance of all life. Shall we not connect the soil more directly with nutrition as we study and come to appreciate these larger national patterns?

Plant nutrient elements and compounds adsorbed on and exchangeable from the soil colloids—not necessarily in aqueous solution—are the fertility "available" to the plant roots.

The clays, that is, the colloidal fraction of the soil—are the seat of these chemo-dynamics. The nutrient elements are adsorbed there from the solutions of fertilizers, for example. They are moved from there to the plant roots. Such elements in their soluble forms do not move into the plant root along with the water necessarily because the fertilizer salts containing them are soluble, or so-called "available." Water movement into the root follows one set of laws; nutrient movement follows another set.

When the clay fraction—a negatively-charged colloid—is highly stocked with adsorbed fertility elements, then the soil is nearly neutral in reaction—or not very acid. More acidity in a soil means merely less fertility per unit of colloid.

Increasing acidity in our natural soil is, therefore, simply decreasing fertility. The presence of the hydrogen is not the detriment. Instead, some hydrogen, or some acidity, is a benefit in connection with fertility. It "mobilizes" the nutrient cations. It will help the acid clays in their "processing" of the reserve minerals to make them become more "available." It helps some soils to "recuperate" when we "give them a rest."

The acidity, or hydrogen, resulting from the respiration of the root, exchanges itself to the clay for other cations—that is, positively-charged ions—of nutrient service to the plants. The acid on the clay breaks down the reserve minerals to move more nutrient ions from them to the clay, and from it on to the roots. Acidity, then, is the chemo-dynamic power with the soil (but of sunshine origin above the soil) to keep the fertility assembly line of agricultural production running for full output. It does not run because of the water soluble fertilizers there.

SOIL FERTILITY: FIRST CONCERN FOR HUMAN SURVIVAL

More than 40 years ago, when the late Eugene Davenport (1856-1941), then Dean of the Illinois College of Agriculture and Director of the Experiment Station, was about to retire "after 34 years of service as a world-recognized authority on farm economics," he was invited to discuss "the most important agricultural problem, in his estimation, that confronts the human race." In undertaking that responsibility, he asked these questions in advance.

"Is it possible so to operate the land that it shall yield profitable crops and at the same time increase, instead of decrease, its store of fertility, or must humanity exist at the expense of the natural supply of fertility as it does of coal, for example? Are we, as people, simply going through a brief geological period, only to become extinct like species that have gone before, as soon as we have destroyed the conditions of our own existence, by working out the stuff of which food is made? Or may we hope for perpetual productivity?"

In his paper, following soon after, he took up those questions on the maintenance of productivity of our soils. He discussed them under the title, *Will the Human Race Finally Starve?*

"Briefly put, the strongest impression on my mind, today," Dean Davenport said in *Farm and Fireside*. volume 46, October 1922, "in respect to agriculture is that, nationally, we have struck a problem for which no solution as yet appears, and from which, unless solved before many generations go by, nothing but national decline seems possible."

"Now, as an economic proposition, there will always be the average man and below. We may raise the average, but it will always be the average. Financially speaking, the average man will not be well-to-do; but scientifically he could be, if he would take care of the land properly.

"Of course, if he could so raise the average as to include the protection of the land in his scheme of things, the public would be safe even if the individual did remain on the dead level of mediocrity. But it is extremely difficult to convince this average man that he should exercise his ingenuity, run into debt for fertilizer, conduct a more complicated and therefore more dangerous business, unless he can be shown that he will profit by the extra labor and hazard. If he does so profit, then he will no longer be an average man, but will rise above it; and if he does not, then he will revert to the easier and cheaper method, which is one of fertility mining and soil depletion."

"So there we have it. A few of the better farmers are so adding to the soil that they are maintaining its full stock of fertility, and will leave the land better than they found it. However, not only the average and below, but also a good many above the average, are mining the land as effectually to exhaustion as we are working out our coal mines and as we have destroyed our great forests.

"The hopelessness of the situation is the depressing part of it, as we consider this man below the average in whose hands and under whose management so large a share of our national lands must rest.

"Frankly, I have no solution for the problem, though a solution must be found. I would advocate a national subsidy for land improvement as for the merchant marine, were such a thing practicable or possible . . . It would be better than going the way of the Pharaohs . . ."

"We are very rapidly building up a great country, very largely at the expense of the virgin fertility of our lands. A few farmers here and there are applying fertilizers intelligently; but considered as a national enterprise, the enormous crops which we are taking out of our land from year to year are so many subtractions from our native store of wealth, which is the free gift of nature."

It is significant that when we ponder the exploitation of soil and the agricultural problem now, 40 years later, we are compelled to agree with the Dean's remarks and to confess that we apparently have not yet found a solution. As we catalog the several degenerations of the body, called diseases, we can do no better than to raise the same question, "Will the human race finally starve?"

RULE OF RETURN

The maintenance of the organic matter of the soil as a requisite for British agriculture was brought to public attention in 1940 under the slogan, "Rule of Return." This was the keynote for seven authors, distinguished men in their respective varied vocations, aiming to bring into clearer focus the national importance of the soil in wartime.

Their pronouncements, so important for American agriculture at the present peacetime, were assembled under the editorial direction of an eighth able writer on agriculture, H.C. Massingham. Their offerings resulted in a 150 page book entitled *England and the Farmer*.

The remarks hidden in libraries deserve repetition to emphasize the role of organic matter in our soils as the chemical buffer which maintains the nutritional quality of the food we grow. They might constitute helpful reading under the title, *United States and the Farmer*.

"Even today," the editor says in his introduction to *England and the Farmer,* there exists a coherent and articulate body of opinion in the country which opens up a unanimous remedy for the appalling ills of our farming lands and most weightily expounds that remedy in term, plant, and proposition opposed to the specialized and central economic-mechanical system imposed by the town upon the country.

"One of the writers mentions a useful shorthand term for that way out—the 'Rule of Return'—and its meaning carries a world of new-old values upon its back. In the most primal sense it implies the return of all waste products to the soil, wherein myriads of microorganisms convert them into the one indispensable substance for every type of plant growth: humus. Without humus the earth becomes a corpse, as the Gobi Desert or the Sahara is, and the enormous increase of desert condition since the development of 'scientific' land-forcing based on finance is a phenomenon even more terrifying than the savagery of the present war, which is the logical outcome of living for wealth rather than for health. How trifling this first necessity of the security of man appears!"

"The act of recruiting death to serve the needs of life restores the balances and rhythm of fundamental life, dislocated by our fantastic idea of 'conquering' nature. By making proper use of this waste, we cut the heart out of another kind of waste: the waste of the earth's fertility by which alone we can live; the waste of health involved in relying upon imported and preserved foods; the waste of resources in the system of usury involved in living upon the forced products of distant lands instead of the natural and nutritive riches of our own; the waste in the advertisement and distribution of dope-artificials; the spiritual waste of separating man from the earth that sustains, renews, satisfies and invigorates him.

"The 'Rule of Return' also rehabilitates the owner-occupier, small-scale farmer whom the treatment of the land, as an industry controlled by big business and on a footing with urban industries, has almost expelled from his native and deeply cherished soil."

"The self-contained and self-sufficient farm . . . allows for the use of machinery to a limited degree; what it prevents is the automatism to which the industrialization of modern Europe has so nearly brought us. It is the small farmer who alone can restore the concept of the family to its prime significance in the life of man and can rescue individual responsibility from its drift to the grave. It counters the evils of mass-production and revives an intimacy with the soil which sees it as a living thing rather than mere ground to be run like a factory and exploited to its utmost extent of productivity."

More than two decades ago the significance of organic matter in the soil and in a healthy agriculture was so well presented in England. Among us in the United States there are increasing numbers of spokesmen calling attention to the declining humus in our soil. They are reminding us that, for healthy life forms, the required protein is more than just the chemical element nitrogen; and the services of the soil's organic matter in plant nutrition constitutes more than a delivery of this single element in the equivalent of what is contained in a bag of commercial fertilizer salts. The application of the latter does not substitute for nature's "Rule of Return" which she allows in growing her climax crops.

MAKE TAX ALLOWANCE
FOR FERTILITY DEPLETION

The 85% of us living in urban areas do not yet feel any obligation to help maintain the fertility resources coming to us gratis from the people in the rural areas.

How soon will we wake up to the obligation we owe to those who maintain reasonable levels of soil fertility so that we may be fed well?

We are set up in urban commercial businesses and industries of which the laws, economics and taxation procedures are so formulated under carefully lobbied legislation that our capital investments in them are self-perpetuating. Even for the minerals or rocks taken out of the limestone quarry, for example, the owner-investor may be allowed a depreciation, or depletion, figure as high as 15% of the income. For the owner-investor in an oil well, it may be a larger amount. The capital investment in these mineral businesses is soon recovered.

But for the mineral fertility taken out of the soil and delivered in the crops to the urban population without charge for it, there is as yet no economist or authority on taxation suggesting the justice of a depletion allowance to the landowner, or investor in that kind of real estate, for the perpetuation of his soil fertility capital in his farming business.

His investment in the minerals in the soil for the food production for all of us is being liquidated gradually under the economic thinking (or the lack of it) which contends that the farmer is thereby taking a profit. On the contrary, he is compelled to throw his financial, and our national security by installments into the bargain every time he makes a sale of his products. Those of us on the urban receiving end of that transaction get those installments gratis and flush them into the sea.

We are parties to the crime of soil fertility exploitation, but yet are crying against the rising costs of living. We are slow to see that such short-sightedness in our economic, agricultural, and other policies toward the fertility resources in the soil are undermining seriously our national security.

All this is the more serious with a growing pressure on the soil's production potential by our own increasing population to say nothing of that by the rest of the world calling on us to share that potential with them.

Liming our soils deserves consideration as an operation undergirding our future security in food, and particularly those foods of high protein content.

We have long known that lime is needed for legumes. We are slow to see that need as one for the production of the protein, rather than the tonnage yield of the crop. It is lime via that route that gets us our meat, milk and eggs. Viewed in this light, one cannot escape the question whether we dare expect the farmer to continue liquidating his fertility assets under the false concept of taking a profit and at the same time ask him to purchase large amounts of calcium and magnesium to aggravate his rate of liquidation all the more.

Isn't it about time that as a basic agricultural policy we design the required machinery of economics and taxation to guarantee the self-perpetuation of the farmer's fertility capital which must feed all of us, both urban and rural?

Perhaps now that the fertility restoration by liming the soil is moving itself into the more exact category of soil chemistry for the nutrition of our plants, our animals and ourselves—should not the maintenance of the soil fertility and thereby of agricultural industry be interpreted by the same views in economics and taxation as those prevailing in other industries?

Perhaps we can bring about self-perpetuation of our soil fertility capital under the agricultural business in the rural areas in the same manner as perpetuation prevails for monetary capital under all businesses in our urban centers.

If that situation is consummated, then liming the soil for calcium's sake will become big business by meeting the major needs in our soils—namely, lime and other fertility-restoring aids through which there can be guaranteed greater national food security for the future of all of us.

WAR: SOME AGRICULTURAL IMPLICATIONS

Man is a sociable being only when he is well-fed. We start our Red Cross campaigns, church solicitations for building funds and other drives after a lunch or special dinner. When he is hungry, man is only another animal. But he is one equipped to deal more severely with others than are the lower animals. As the old Russian proverb has it, "An empty stomach knows no laws."

Food has always been the major incentive for man's struggles and efforts. In some countries, about 90% of one's time goes to providing the next meal. In others, it takes less than 20%. That struggle is larger as the soil that grows the food is more stubborn; as the food crop becomes harder to grow; or as there are other ways of saying that the creative capacities of the soil in its particular setting are not sufficient to grow plenty of food. That, in simple fundamentals, says that the soil in a particular setting is not fertile enough, or not productive enough.

Now that the world's population has grown to about 2.2 billion people, while the crop land area is only about 2.4 billion acres, we each count our share down to but a little more than an acre apiece. We are beginning to visualize not only our own but also the international situation as a problem of providing food.

Perhaps you have never considered the international implications of the facts (a) that there are no significant numbers of unoccupied acres to be put into production as additional food accommodations for the growing population, (b) that the declining fertility in the acres long under cultivation and erosion is shrinking the food output in tonnages per person, and (c) that food quality as protein, in meat, milk and eggs for example, is the real problem now so acute because the soil fertility required to produce these protein products has gone too low. When on much of our land it requires two acres even to keep a milk cow, can you see one acre apiece giving each of us enough proteins?

Proteins are the food constituents that grow the body. Carbohydrates give us only energy, or make fat. Life flows through the proteins. They can multiply the cells or make for growth. They make the plants' seed and represent the processes of reproduction. Proteins as food are both growth and energy. Fats, like the carbohydrates, are energy only. Proteins are blood; they are muscle; they are strength; and they are protection within the body against disease. They are not created from the elements by our animals. They are so created only by the plants and the microbes. Then, from those parts the animals assemble them into more complete values of nutrition for us as higher animals.

Only fertile soils can make seeds or create sufficient proteins in those concentrated forms. Any plant is first vegetative bulk like carbohydrates which we too readily call plant growth and crop yields. Only its seed represents the concentrated protein.

Failure to make seed is failure to make protein because the fertility of the soil was failing. Vegetative mass worked over by the cow and her microbial cohorts in her paunch and intestines may yield some proteins. She gets much more, though, per mouthful from eating the grains. As for ourselves, we must depend on concentrated proteins in the grains. Humans fed directly by crops must have soils fertile enough to let the plants make the grains. Vegetation failing to make seeds on infertile soil is commonly considered "cow feed." While cows can use vegetative masses in volumes beyond us, even that must be more than bulk and the soils growing it must be fertile enough to give her the necessary proteins, minerals, and other soil-borne substances connected with the creation of proteins.

The problem of proteins is not solved by passing it to the cow. It goes back to the soil by which we too must be fed. It is basic to our international situations in ways not caused or controlled by politics.

Present turbulent politics are merely the symptoms of what has seemingly its major cause in the insufficient protein from insufficiently fertile soils. It is centered namely around the desperate struggle for proteins.

A study of the protein potentials in the soils of the United States according to the climatic pattern of fertility will be suggestive in our examination of the soils of the world. By that we can catalogue those countries or peoples who have and those who have not protein. Perhaps we can see the soil fertility marking out the groups of haves and have-nots with two of the haves opposing each other from opposite sides of the Atlantic.

A look at the map of the soils of the world after studying the soils map of the United States in relation to protein production, will extend our understanding. It will help us to see that soil fertility in terms of possible output of protein as the food essential in high protein wheat, nutritious grasses, meat and other "grow" foods has some significant implications in our present world situations.

The United States have an interesting pattern of soil development because of the climatic range represented. Our rainfall, for example, is of almost zero quantity just east of the coast range along the Pacific edge. As one goes east from there across the country, the annual rainfall increases to about 40 inches on reaching the Mississippi Valley. It holds to that figure as one goes across the northern half of the eastern United States. But it goes up to 60 inches annually in the southern half of the east.

Since soil development is vitally affected by rainfall or lack of it, we have in this country a variant of soils in all stages of development.

More rainfall on coming eastward means more rock breakdown, more clay and more nutrients on the clay to give, with moderate rainfall the vegetation that makes a deep black soil. These are the areas where the soils haven't leached, nor have they lost their lime content. These soils grow "hard" or high-protein wheat today, but are rapidly shifting over to growing "soft" wheat because of soil depletion. They are astride the line that divides the United States, putting the less-leached soils to the west and the fertility-deficient or acid soils in the east.

It is in the central United States where the soils still retain much of the original rock contents as fertility; where much of this is absorbed on the clay in ready use for the crops; and where much potential food source still remains as less weathered minerals in the silt and sand fractions of the soil. It is in this mid-continental area that legumes can synthesize their high protein contents as good feed for growing the bodies of young livestock. It is there that the plants are active not only in photosynthesis, the process that compounds, the carbohydrates from air and water by means of sunshine power; but plants there are active also in biosynthesis, the process that converts these carbohydrates over into amino acids or proteins by the help of the numerous elements of fertility available in the less-leached soil.

It is in the mid continent where the glacial deposit, once pushed down from the north, was a good mixture of many minerals before the development of soils from that deposit began. It is also the area where the Missouri River in flood times is hauling unweathered silt from the arid West and leaving it as exposed flood plains of low water periods to be blown up and scattered as rejuvenating fertility supplies on much of this more-productive, mid-continental area. It is this combination of many factors that is active in stocking the soil assembly line for building proteins.

With the exhaustion of soil fertility either by man's use of it in cropping and crop removal or by nature's excessive leaching under heavy rainfall, the vegetative output is less of protein and inorganic benefits, but more of carbohydrates. Soil depletion and our readiness to measure production in terms of tonnage only are crowding out those crops that are truly "grow" foods and bringing in those that are only "go" foods. As long as soil depletion was pushing us westward from the Atlantic coast toward the better soils of the mid-continent there was no need for worry. But now when it has pushed us past there and on to the soils that fail us because they are undeveloped, then the national food problems become serious ones.

When the soil map of our country tells us that it is the moderate rainfalls in the temperate zones that develop soil for protein production, we can then use the soil map of the world in contemplating the international situation. A look at the soils of the world points out immediately how limited in many areas of the world are those soils similar to our mid-continent, or our bread basket and our meat basket.

In Europe, the soils duplicating our mid-continent are not in England, or in the western continent. Instead, they are in the eastern Europe or the Russian part. Before the first World War and the destruction of the Czar's family (of German lineage) wheat moved out of Russia via Mediterranean water routes to Germany. Our American wheat exported largely for sale to Germany depended for its price on the Russian wheat crop, the competitor supplying Germany.

After World War I, the Bolsheviks ate their own wheat rather than have the Czar and landlords tax it away from them to balance Russian's international trade, or to have the monks eat and drink (fermented drinks) from it via the church.

Does not a look at the world's map of soils and their protein-producing power suggest reasons why a commander-in-chief of a hungry army, like Mussolini's should try to move into North Africa, or like Hitler, have a push to the east (Drang nach Osten)? Does it suggest that the stealthy move by Japan into North China with its protein-producing soils (soybeans, Manchuria) was a mystery in place of her moving into South China for its carbohydrate-supplying crop (rice)? When a single raw fish head was the daily protein allotment in Japan, are some of the close approaches by those folks to American fishing areas a surprise?

Do we marvel at the mouth of the Uruguay River in South America as the strange place for a sea battle between the German and British Men of War when the center of the war was in Europe? Could food coming out from there have been any reason why warships were active there even when the two countries blockaded and hungry, were on the other side of the Atlantic. Is a decadent British Empire a surprise when her protein supply has always been in the colonies of the Empire, but those colonies are now manufacturing their own goods and fading out as a market where Great Britain can exchange labor for the protein foods her soils cannot produce sufficiently?

Protein-producing soils in the limited areas of the moderate rainfalls in the temperate zones are the objects of struggle in a higher degree than realized when we think no deeper than politics and political leaders. We forget that politics are an expression of a more deeply seated cause. It may be well to view the politician as one trying to find which way the hungry crowd is pushing so that he can get in front to be pushed in first. Can it be helpful to view the international tensions today as pushes by the crowds to get in where the fertility of the soil is still high enough to make proteins, the foods that truly feed rather than only fill and fool us?

On the basis of the soils and their fertility, there seem to be some sound foundations on which to build our thinking about threatening situations. Perhaps some help can be seen in viewing our own soils as their fertility has far reaching implications, even to international dimensions.

THE FRENCH DON'T DARE WEAR OUT THEIR FARMS

The request to teach soils to our American soldiers in the Biarritz-American University of France, brought with it opportunity to study the soils of France as the basis of an old agriculture.

Teaching soils was an international affair. It certainly was extensive in its geographic aspects, with a class in soils including students from 38 of the United States, France with her colonies, Great Britain with her outlying possessions and in fact, any people who could understand the English language. I would have been neglecting an excellent opportunity in this transplanted college of agriculture, if I had not taught most objectively by using the soils, crops, livestock and other agricultural objects right before us in France to illustrate the fundamental principles of the subject in the classroom.

In this older country and its older agriculture, one soon recognizes the adjustments of this means of livelihood to the soil as this determines the nature and magnitude of farming. These adjustments clearly demonstrate what is meant by saying "as one gets older he becomes more conservative." This old agriculture is very conservative. Conservation of soil fertility is easily recognized and forcefully imprints itself on one's mind.

Livestock plays a big part in putting back on the land much of the necessary chemical elements or fertility supplied by the soil. Farm animals have been doing much to return soil fertility to keep life over here going during the many past centuries.

In looking at France, and most any other of these older countries of Europe, our soldier boys have commonly made these comments: "The farmers of Europe seem terribly slow. Farming with oxen as power certainly doesn't consider time." "Those farmers certainly put a lot of work into growing and harvesting their crops."

Then one also could hear the soldier-students say: "But they certainly love their animals, if one can judge by the way they care for them." "They certainly don't waste anything over here, not even the manure. The way they fuss with it makes it look as if it were something very valuable. Perhaps our own agriculture in the United States is heading rapidly toward this European kind of farming if we don't start practicing conservation of our soils and other resources more completely."

The sojourn and study by our own farm boys as soldiers in Europe predicts a wholesome future effect on our farming. It may be inducement for a shift from a highly exploitive system to one that is much more conservative. Our boys realize that the folks in this country who have learned to conserve have had the means by which to maintain themselves well into old age. The agriculture of this country, which the Romans called Gaul long before the Christian era, already has lasted 20 centuries because the conservative use of the land has left its soil with enough producing power to be a support for agriculture today. Families keep their land and pass it on down through successive generations of their offspring. These folks keep the land; they keep the soil; and the soil in turn keeps them.

Livestock is an integral part of the farm living. It is not just a speculative aspect of the farming business. Cattle are not grown rapidly in large numbers and sent off to slaughter. This would mean a heavy drain on the soil fertility in terms of calcium and phosphorus in the bones. It would also draw heavily on the nitrogen in the soil needed to grow the muscle. Such high rates of fertility removal in the past would have been too highly exploitive for the soils to last so long.

The French farmer does not grow ten head of cattle in one year, sell them off and forget the soil fertility he is throwing in with the bargain. He does not have what might be called ten cow-years telescoped into one year. Rather, he grows and keeps one head ten years.

In that practice one may well look for the necessity that compels it. It is a part of the balance, or the adjustment, of the agriculture by which the soil can continuously supply the fertility. French farmers dare not wear out their farms. Their country has no more new farms for them.

While the French farmer's one cow is putting out ten single cow-year units by living ten years, which we would get from ten baby beeves in one year, she is occupying a prominent place as a regular piece of farm equipment and as part of the very farm itself. She is serving as a triple-purpose animal. She is the farm power. she provides milk, though perhaps not in daily quantities we might deem necessary. Eventually she is meat. In serving the last of these three purposes, one might see in her extended age some explanation as to why her carcass requires some mechanical aids at mastication. That's why it comes to the table largely in the form of sausage. We see why meat requires long cooking in water to end up in soup—which is the national dish introducing most dinners in France.

The mature cow is much more efficient as a conserver of soil fertility than the growing calf. The nitrogen, calcium, phosphorus, potassium and other fertility elements going into her as feed are only body replacements. Their equivalents are put out as urine and feces that represent a byproduct with fertilizer possibilities coming along with her three-phased services.

The young animal hauls these nutrients off the farm. The mature cow serves mainly to rotate them there with a small loss while she is burning their organic combinations to provide power. That power represents air, water and sunshine concentrated into usable form by means of this soil fertility. The latter is not hauled off, but can go back via the manure into another cycle of oft-repeated services.

Conservation of the fertility has been the habit with these folks. It is not a new practice encouraged by recent special instructions and appropriations. One may then well raise the question whether any agriculture would have survived these many centuries if conservation of the soil fertility had not been its regular habit.

One needs only to see the French farmer's attention to the bedding and the manure to recognize conservation of soil fertility in practice. In areas where straw is not plentiful, he brings in bracken fern from the forests, reeds from the summer-dry swamps, and the lower parts of the cornstalks from the field. From the last of these, the tops, the lower leaves and the ears have already gone from the field as feed. But this coarse bedding would not readily soak up the urine, considered too valuable to be allowed to run through and be lost. Consequently forest leaves are mixed in. Were the leaves used alone they would laminate to be so compact as to turn off the urine for its loss. But when mixed with the coarser materials in the proper proportions as learned from experience, they help to retain the urine while the tramping breaks the stalks and opens them for its better absorption. Full well does the farmer know his bedding materials in terms of their soil fertility values, when he discriminates between the different kinds of forest leaves in telling you that one kind does or does not make good manure.

The French farmer truly "makes manure," and—quite unlike our own practice—does not just haul out for disposition what is a barn waste consisting of bedding and animal feces. For him such materials, as he takes them out of the barn, including the carefully conserved urine, are not manure as he speaks of it. They are merely the raw materials which he puts up carefully into a straight-sided flat-topped pile in order to let them go through the heating process. This process is, for him, the making of the manure, which means production of an organic fertilizer. Perhaps he cannot give you the chemical and microbial techniques involved in the process, but he does not saturate the pile initially with water and bring on exclusion of the air. Instead, he uses the woody, carbonaceous bedding as the energy foods or fuels for the decay-promoting microbes, just as we feed starchy grains or carbohydrates as energy-foods for livestock. Just as we balance the grains with protein supplements, so the urine, carefully conserved, supplies the nitrogen as supplement to balance the carbon in the microbial diet from which the excess carbon is respired or burned by the growing microbes to the extent of heating the manure pile.

Soluble minerals come along with the nitrogen in the urine. They improve the diet of the bacteria that convert this original mixture of coarse bedding into a microbial product that serves well as a fertilizer for crops. Once the heating is finished, he keeps the manure pile nearly saturated with water to exclude the air and stop further respiration or the burning of this finished product. He may put it out in the field in regular rows of uniformly-sized piles in order that the reduced temperature may hold down the rate of its destruction, and from which he can distribute this fertilizer at uniform rates.

Although the farmer of France may speak of this byproduct of his power animals as manure that he has made, it can well be considered a fertilizer which he has manufactured from the barn wastes by proper microbial management. Strawy manure plowed under for a crop like corn—as some farmers have learned from costly experience—makes the crop yellow and of sickly appearance, or "burn it out" as is commonly said. The manure made into a fertilizer in the same sense as the European farmer makes it avoids this danger. Piled, as it was, it has already burned out the surplus carbon while the rest of it was combined with the nitrogen and other solubles in the urine. He does not feed this extra woody material to the soil microbes. They are, therefore, not using the soil's supply of soluble nitrogen to balance the excessive carbon in their diet, and are not leaving the competing corn plant in that starved condition for nitrogen which is so often wrongly considered due to a shortage of water.

Instead, he plows under an organic matter from which the microbes must get energy, but in which there is carbon in a narrower ratio to the nitrogen than they normally use so that they get some nitrogen free for the plants instead of competing with the crop for this nutrient. We must admit that he knows his carbon-nitrogen ration for his soil microbes just as we believe we know our nutritive ratios for putting feed through our livestock. We must admit, also, that we can well extend our better understanding of the conservation of soil fertility in farm manures by studying the practices in these older countries.

Yes, the European farmer knows how to "make manure." His farm power coming by way of livestock also is a good exhibition of the practice of conservation of soil fertility developed to a high degree. This conservation is not only a matter of putting the chemical elements back into the soil as we would recognize them in the ash after the chemist's ignition and analyses. Instead they are going back in complex compounds of organic matter. These may be much more than just so many elements as we readily calculate them in recommending a formula for commercial fertilizers. They may be the plant hormones. They may be many other complexes of values yet unknown, but like vitamins tremendously important even if needed in only very small quantities.

The European farmer who still persists in making manure and thereby making his organic fertilizer, may be practicing a kind of conservation that is more far-reaching than those of us who are using our soil fertility so exploitively are likely to appreciate. In the United States our agriculture is still young. But yet, we already have abandoned and ruined by erosion extensive land areas. Are not those eye-sores reminding us that we must take to being more conservative of our soil fertility if we are to continue in agriculture; and that we must do so long before we get to the agricultural age of countries like France which we may—possibly derisively—consider "old and conservative?"

THE SOIL AS A FARM COMMODITY OR A FACTORY

It was only a few hundred years ago when the nomadic pastoral existence of peoples shifted to a stationary agriculture and brought with it territorial claims in the name of the individual who tills the soil. That period gave us land ownership with full land and mineral rights, and complete freedom as to land use. It gave birth to the commodity concept of the land, with privilege of barter or sale of the land area in question.

Viewing the soil as if it were a common agricultural commodity that we might sell, or save, we have brought ourselves face to face with the need for its conservation, not only in territorial area and body mass, but in terms of the producing capacity inherent in its chemical composition as plant producing value.

Now that land has been bartered and traded for years, not only its space dimensions have been carefully and legally defined, but its productivity is undergoing refined definition. Under that attention land has become soil, as a distinct object of study and classification. Values formerly unknown are being attached to it and magnified. Emphasis on these values of the soil is doing much to move land out of the commodity picture or concept, and, as in older countries, to take it off the list of items for simple exchange and sale. We shall be moving rapidly toward a higher appreciation of the soil as it represents our future security and less and less as it is to be considered a commodity.

When carefully examined for its services and functions, the soil plays a dual role. In the first place it is a stock of raw materials. In the second place, it is a factory. As the former, it must contain and retain in stable form against loss by weathering, those items necessary for annual plant growths over an extended time period. As the latter, it must convert the more resistant plant nutrient elements into such forms as are taken by the plants and at such rates as will produce significant plant bulk for an economical crop. We can therefore not view the soil as if it were so many pounds of a commodity to be sold like we sell butter, beef, or grains at so much a pound. Rather the soil is the locality where under sunshine power, the contribution of but 5-10% of mineral or ash materials by the soil starts their assemblage with the other 90 to 95% as inflow of air and water to give us that mystical performance we call crop growth, or agricultural production, and the greatest creation of wealth for human good. Considering the matter from that viewpoint no one would deny the need for soil conservation. Who would destroy the very tool of the earth by which the sun and the skies may be brought to clothe her and to feed and shelter her people?

Because of a limited understanding of the functions of the soil, we have not had a clear concept of handling it even as a commodity. We have not known, or reminded ourselves, what part of it is plant nutrient, much less how much of each nutrient supply we are selling annually. We have not translated our stock of goods on the shelves in the soil store into monetary units which can be added or subtracted, or can take in part an inventory. Each deal, or sale, has not made the fertility cash register ring up the removals so as to give us totals at the end of the day or year. We have not translated soil organic matter, phosphorus content, calcium deficiency, or soil acidity, into units of human effort and deeper concern, by which values are most permanently fixed and appreciations most effectively generated in our minds. Given to us by nature for the grabbing (still the method of territorial procurement by many nations), we are yet to develop the appreciation for them, and seemingly only after their exhaustion.

As a partial help toward understanding the soil's contribution to plant growth and the magnitude of the performances involved in making a crop, let us first turn to a picture of the composition of vegetation in general for the soil's part in it. The soil contributes but 50 pounds per 1,000 pounds dry matter. But unless these are delivered promptly, the remaining 950 pounds will not be captured from the air and water, and the sunshine power will remain unused.

Conservation of the soil in the fullest sense does not call for the hoarding of the plant nutrients in the soil, but for their wisest use with fullest preservation within the soil. Using more familiar cropping systems, or specific rather than general vegetation as given, we may well examine our business of selling the soil to learn how long we can continue marketing it as a commodity. Much better, let us try to learn how we can run it as a factory without selling off annually much of its essential equipment.

Commercial fertilizers have come to give us a monetary value of plant nutrient items. It will be helpful to determine what the annual cash register report would be, should it ring up the fertility removal in terms of the respective nutrients, at fertilizer prices for their return. It may bring appreciation of the fact that the annual fertility turnover is an unrecognized "big business." In that turnover there are hidden some secret costs and some bookkeeping that is deceptive to degrees more challenging than many accounts submitted to congressional investigating committees.

As a simple case, let us assume a 200 acre farm on virgin Marshall silt loam divided into five 40 acre units, bearing, corn, oat, wheat, clover, and grass.

With a subsoil laden with storage water and with liberal yields per acre, no larger than 50 bushels of corn, 35 bushels of oats, 20 bushels of wheat, 2 tons of clover hay, and 1.5 tons of grass hay equivalent, a year year round on an acre would remove the following in the rotation annually and with the following monetary values (see chart on page 268).

Perhaps it will be no mistake if we disregard the values of the iron which is required in such small amounts as contrasted to the supply in the soil; or if we assign no value and no detriment to the sodium and the silicon, respectively, both considered non-essential for plant growth. Before dismissing the significance of sodium too light-heartedly, however, it may be well to remind ourselves that it was only about one hundred years ago that the sodium and chlorine delivery by crops dropped to the low level that made agricultural discussion common then as to whether it would be good practice to feed animals ordinary salt or sodium chloride. Then too, our declining soil fertility is reflecting an increasing share of the plant ash as silicon as we go from the vegetation on the more productive soils to those less productive, or as we go from the better legumes of higher feeding value through the plant series to non-legumes and lower nutritional significance.

The significant item in the table is the high value of the fertility taken from the soil annually, given as the figure, $8.15 per acre. Have you ever imagined that on a 200 acre farm the fertility equivalent of $1,630 is taken out of the soil annually? We do not appreciate the possibility for economic and soil fertility leaks that this unrecognized capital turnover involves. In this capital turnover and its leaks, there are well smothered most of the economic disturbances for which the numerous remedies in the form of changing gold standards and other panaceas are so freely offered in our thinking which stays at no small distance above the surface of the soil. This $8.15 an acre represents an annual production of 3,824 pounds of organic output at the cost of 200 pounds of the soil commodity. One pound of the soil is expended per 19 pounds of produce grown. In terms of the fertilizer prices for the soil expended, the organic matter produce is costing more than 20 cents per hundred weight.

Have we granted that every ton of produce coming from the field is liquidating $4.00 per acre of the assets we initially purchased in the form of a farm? Have we thought that a rent of $4.00 per acre would cover only one-half of the fertility coming out of the soil? Who has been willing to grant that in our economic arrangements we have failed to charge against the produce going off the land this cost to replace to the soil the removed fertility, even if we handle our soil only as a commodity? We have been selling our soil without appreciating its cost, and then find ourselves at a loss to understand our economic troubles. We haven't even understood the material we are selling as a commodity.

Should we view the soil more as a factory and less as a commodity, we would return more of the stuffs taken from the soil store and would market more of that portion of the produce captured from air and water. Let us examine the idea of passing the vegetative produce through the animal machine to learn how much the animal will carry away and how much can be returned to the soil.

The working or fattening animal excretes, as an average, 85% of the fertility ingested. Then if the $1,630 worth of fertility capital all goes through such animals they would excrete $1,385. Should they be milking or growing animals a 65% recovery would excrete but $1,060 of fertility that might go back to the soil. Even with animal use of much from the soil and with no manure loss, one must still be prepared to write off an annual inventory loss of soil fertility on this farm amounting to $245 and $570 annually in these two cases.

Such figures take into account no loss of the fertility in the manure before getting it back to the land and are only the losses going off "on the hoof." Additional losses must still be proportionate to the difficulty or the carelessness in manure handling or getting its fertility back to that part of the farm where crops can use it again. Should this manure loss be held to the minimum of about 25%, then this would permit the animal fattener to return $1,222 or 65%, and the dairy farmer $795 or 50%, of the $1,630 of the rotating fertility capital. Should they permit one-third loss of the manure, which still represents good manure management, then $923 and $706 or approximately 45% and 55%, respectively, of the fertility capital are taken away from the soil.

Should any grain or forage be sold rather than be fed, then by just these percentages of that item's fertility content would the rotating capital suffer additional loss, and the soil be that much more nearly in the commodity sale category rather than in that of the factory.

It might seem that fruit farming, where so much of the bulk of output is water, should be more conservative of soil fertility. Let us take an apple orchard of larger tree size with 30 trees per acre, 40 feet apart. In terms of only the three major plant nutrients, the total, annual rotating fertility capital would be represented by $2.74 per acre. This, for the apple grower, is much less than the $8.15, which was the amount for the general farmer. But again the loss is 55% of this amount. Though smaller in the absolute, it still represents a loss of more than one-half of the offerings by the soil.

For the peach grower with smaller trees, at 100 per acre, the balance sheet would show a total annual rotating capital of $1,306 per acre. In his case the removal is greater in absolute than for the apple grower, or for the general farmer and again 45% of the rotating capital goes off the soil.

Thus fruit farming in apples is lower, and in peaches higher in rotating capital annually per acre. The fruit farming does not lessen the responsibility to the soil, only perchance as the seasonal risk is greater and the fruit farmer does not so nearly duplicate the figures annually.

It is conceivable that there may be methods of producing farm products of such nature, and of marketing the parts of such chemical composition so that little more than carbon, hydrogen, and oxygen originating in the air and water is moved off the farm. Should only butter go off for example, it would sell no soil fertility save that in the casein and in solution in the water in the butter. Sugarcane juice is seemingly mainly carbohydrate, but yet even it carries fertility along with the sugar in solution, reflecting in the juice the fertility of the soil growing it.

Escape from the problem of returning fertility is not an easy one. It may be more simple to accept the fact that almost any system of farming brings with it the responsibility of knowing how much fertility is leaking out of the rotating supply and how much must be regularly restored.

Should we take the simple calculation that about 50% of the rotating fertility is being removed annually, it is still necessary to recognize how this is related to the entire stock in the soil. Using Marshall silt loam with 3,800 pounds nitrogen per acre in virgin surface soil producing the crops previously listed with yields not above 50 bushels of corn or other crop equivalents, with the loss—not covered by clover return—amounting to 19.1 pounds per acre per year, this would be just one-half of 1% of the total. Should these annual rates of loss continue, then the Marshall silt loam would be exhausted of its nitrogen store in 200 years. Using the soil as a commodity it would be "sold out" in that time.

More complicated calculations are needed to express the declining curve accurately, but yet these results are sufficient to bring us to the conviction that our soils can not be taken as commodities—and still serve as the basis of a permanent agriculture. If our farming is to be permanent and profitable, our soil must be a factory to which we bring regularly or restore promptly the calcium, phosphorus, nitrogen, and other fertility elements escaping from the stock of rotating capital, and in which we maintain these at high operating level. We cannot have permanent agriculture that sells out its soil as a commodity, nor can a national economy of permanence be built on an agriculture operating blindly on such a policy toward its soil.

DOUBLE CROPPING FOR DOUBLE PROFITS ... REQUIRES GOOD SOIL MANAGEMENT

The high cost of seedbed preparation has led to the practice of growing the cereal nurse crop-legume combination in order to make the one preparation serve for two or more crops. Increasing soil erosion has also stimulated the desire to reduce cultivation to the minimum and keep soils under crop cover the year around.

Double cropping systems readily present themselves as a hope for better soil conservation and for double profits. For this service, however, more than one kind of crop is required. Failures of legume crops have shown that in many sections seeding of these soil improvers are more often successful when (1) they are made without a nurse crop; (2) the legume follows rather than accompanies the nurse crop and (3) calcium, lime or other fertilizing materials are used as soil treatment in advance of the combination seedings.

This good soil management calls for the best annual seedbed preparation by plowing and judicious application of extra plant nourishment in limestone and other fertilizers.

Plowing improves the structural arrangement of clay and the other larger mineral crystals that make up the soil. Plants pick up nourishment more readily from clay than from silt, or the larger mineral crystals, but the latter is more nourishing to begin with. Hence the problem of feeding the plant for better growth and its higher nutritive value as animal feed is one of getting the nutrients transferred from the silt to the clay so that plants get them more easily. This transfer is speeded up by shifting the connections between the surfaces of the particles of the silt and the clay. Plowing comes in the "pass the ammunition" from the soil to fill the plants with proteins, minerals, vitamins and other growth promoting substances. One must invest in plowing and fertilizing if he is to get double profits from double cropping.

To further the all-out war production effort many farmers are planting buckwheat, or soybeans for hay, after early maturing small grain is harvested. Some are plowing up old meadows for such crops as these, after the hay crop is put up. Still others are cutting first year sweet clover for hay instead of allowing it to remain in the field to be plowed down for soil improvement the following spring.

Various other cropping combinations have been under trial. Wheat, oats, barley and rye have been used as the cereal crops in combination with lespedeza and soybeans.

There are no unusual virtues in the crops or their combinations that guarantee the profits unless the best of soil management supports them with liberal supplies of nourishment. Crop production is measured by the fertility of the soil. Double cropping can give double profits only if the soil guarantees such a return.

If the seedbed has been properly plowed, if there has been intelligent application of limestone and fertilizers; if crops have been well chosen, double cropping will bring double profits. Now, while the need for food is great, is really the time for intensive liming and fertilization, and intensive cropping. This will benefit the nation as well as the farmers.

SOURCES

CHAPTER 1. THE NITROGEN FACTOR

PAGE	NOTE

1   Is Nitrogen Going West? reprint from *The Fertilizer Review*, January-February-March, 1946.

5   Is Commercial Urea an "Organic" Fertilizer? *Let's Live*, August 1956.

6   An Old Problem—Loss of Applied Nitrogen, *Let's Live*, December 1961.
      1. Agricultural Chemistry. Printed for the Society for Promoting Christian Knowledge, London, 1849.

7   2. Liebig, Justus von. *Die Chemie in ihrer Anwendung auf Agrikultur und Physiologie.* 9th Ed. p. 27.

8   Nitrates Possible Poison Grown Into Foods, *Let's Live*, July 1961.

9   1. Garner, George B., "Learn to Live with Nitrates." *Mo. Agr. Expt. Sta. Bul.* 708, June 1958.
      2. Mayo, N.S., "Cattle Poisoning by Nitrate of Potash." *Kansas Agr. Sta. Bul.* 49, June 1895.
      3. Wilson, J.K., "Nitrate in Plants: Its relation to fertilizer injury, changes during silage-making and indirect toxicity to animals." *Jour. Amer. Soc. Agr.*, 35:279-290, 1943.

10   4. Gilbert, C.S. et al, "Nitrate Accumulations in Cultivated Plants and Weeds." *Wyo. Agr. Exp. Sta. Bul.* 277:1-37, 1946.

11   5. Heart, J.G., "New 'Blue Baby' Diseases." *Science News Letter LIII*:number 18, 275-276, 1948.

12   Soil Nitrogen Still a Big Problem, *Let's Live*, June 1950.

13   The Role of Nitrogen, *Let's Live*, February 1953.

15   Soils Need 'Living' Fertility! *Western Livestock Journal*, January 1963.

17   Nitrogen for Proteins and Protection Against Disease, *Victory Farm Forum*, September 1949.

22   Declining Nitrate Levels in Putnam Silt Loam, reprint from *Journal of the American Society of Agronomy*, volume 26, number 7, July 1934.

27   Fertilizing with Nitrogen: The Cow Makes Her Suggestions, *Let's Live*, June 1957.

29   Fertilizing with Nitrogen: Rabbits Testify by Experiments, *Let's Live*, July 1957.

2. "A mind free from impressions. A scraped tablet."

CHAPTER 4. THE ANATOMY OF SOIL FERTILITY

CHAPTER 5. PASTURES

Agron. 31:284-286, 1939 (number 4).

4. Hampton, H.E. and Wm.A. Albrecht, "Nodulation Modifies Nutrient Intake from Colloidal Clay by Soybeans," *Proc. Soil Sci. Soc. Amer.* 8:234-237, 1944.

204 Soil Fertility and the Human Species, reprint from *Chemical and Engineering News*, volume 21, number 4, February 25, 1943.

215 Blood Will Tell, *Let's Live,* November 1957.

217 Is Soil Fertility Via Food Quality Reported in Your Varied Pulse Rate? *Let's Live,* February 1957.

218 High Time to Learn About our Soils and Our Health, *Agricultural Leaders' Digest*, June 1953.

220 Health Depends on Soil, *The Land,* May-June-July 1942.

227 Our Teeth and Our Soils, reprint from *Let's Live,* volume 2, number 2, May 1950.

229 Bio-assays Rather than Chemical Analyses, *Let's Live,* October 1962.

1. Sheldon, Victor L., Biosynthesis of Amino Acids by Forage Plants According to Soil Fertility, Thesis, Doctor of Philosophy, Missouri, 1950.

231 Pass-Along Hungers, *Let's Live,* July 1950.

233 Food Quality—as Physiology Demands It, *Let's Live,* October 1956.

235 Nutritional Values of Apples According to Chemical Analysis, *Let's Live*, May 1966.

237 Nutritional Values of Vegetables Change Rapidly, *Let's Live,* March 1966.

239 The Little Things Count in Nutrition, *Let's Live,* March 1954.

241 Nutritional Quality of Vegetables Via Plant Species and Soil Fertility, *Let's Live,* December 1965.

243 Variable Quality Production by Food Plants, *Let's Live,* February 1966.

246 Quality Becomes More Quantitative, *Let's Live,* November 1965.

248 Size and Weight Versus Nutritional Quality, *Let's Live,* July 1966.

250 A New Book—"Our Daily Poison," *Let's Live,* April 1956.

252 Our Soils Affect Nutrition, *Let's Live,* February 1956.

253 Soils, Plants and Nutrition, *Let's Live,* May 1956.

254 Health, as Different Soil Areas Nourish It, *Let's Live,* September-October 1966.

255 1. Nizel, A.E. and R.S. Harris, "Effects of Foods grown in different areas on prevalence of dental caries in hamsters," *Archiv. Biochem.* 26:153-157, 1950.

256 2. Keyes, P.H. *Journal Dental Research*, 23:439, 1944.

3. Albrecht, W.A., "Our Teeth and Our Soils," *Ann. Dent.* 6:199-213, 1947.

257 4. Miler, John J., "The influence of soil minerals upon hu-

CHAPTER 8. PRAGMATIC ALBRECHT

INDEX